Participatory Technology Development

Dr. M. Israel Thomas
Dr. T. Rathakrishnan
Mrs. S.R. Padma

SCIENTIFIC
PUBLISHERS (INDIA)

Published by:

Scientific Publishers (India)
5 A, New Pali Road, P.O. Box 91
Jodhpur 342 001 (India)

Branch Office
Scientific Publishers (India)
4806/24, Ansari Road, Daryaganj
New Delhi - 110 002 (India)

E-mail: info@scientificpub.com
Website: www.scientificpub.com

Print, 2023

ISBN: 978-81-7233-792-6

Typeset by Rajesh Ojha
Printed in India

TAMIL NADU AGRICULTURAL UNIVERSITY

DR. P. MURUGESA BOOPATHI
Vice-Chancellor

COIMBATORE - 641 003
TAMIL NADU
INDIA

FOREWORD

Agriculture is the basis of the economic and social fabric of India. A sound agricultural base alone can sustain prosperity and there by the national development. During the past few decades Indian agriculture has undergone tremendous change, mainly due to concrete research and supporting systems, for developing and diffusing the innovations. The diffusion of innovation is being carried out through mass media, extension agencies and others. However, these channels cannot reach all the farmers. To reach all the farmers effectively Participatory Research is the only means.

Farmer participation in agricultural research is a systematic dialogue between farmers and scientists to solve problems related to agriculture and ultimately increase the impact of agricultural research. In the present day context this approach is very important to increase the efficiency and impact of agricultural research and technology. Through Participatory Technology Development (PTD) approach indigenous knowledge available in the system can be identified and integrated in the modern technology development process.

This book brings together highly rich information on participatory technology development and indigenous technical knowledge by giving due emphasis to experiments and field study in the fields of extension, teaching and research. The authors of this book have made a novel attempt to bring together all the important areas in the field of Agricultural Extension under one rubric. I hope the book will be extremely useful to agricultural extension workers as well as agricultural graduates.

I harmonize the authors of this book entitled "**Participatory Technology Development: A Technique for Refinement of Indigenous Technical Knowledge**" and I am sure this effort will be appreciated both at national level and international level.

(P. Murugesa Boopathi)

● Tel.: Off.: +91 422 2431788 ● Res.: + 91 422 2430887 ● Fax: +91 422 2431672
● Email: vc@tnau.ac.in / pmboopathitnau@yahoo.com
● Website: www.tnau.ac.in / www.agritech.tnau.ac.in

Preface

In developing countries, traditional methods have special advantages over modern agricultural techniques. The capital and technological skill requirement in the use of traditional technologies are generally low and their adoption often requires little restructure of the traditional societies. These traditional technologies are nothing but indigenous knowledge. By adopting such indigenous knowledge, our ancestors did not face any problem of large scale pest out break or economic crisis unlike the today's farmers.

Indigenous knowledge is not confined to tribal groups or the original inhabitants of an area. It is not even confined to rural people. It is dynamic and it changes through indigenous creativity and innovativeness as well as through contact with other system. Indigenous knowledge is available in almost all the agro-climatic zones and in all the farming system.

The traditional practices and their scientific validations were covered in this book and the practices were rationalised. These rational practices will serve the scientists with new technology for test verification and to select the viable technology for field level recommendation.

This book should serve as a valuable tool and provide useful insights to planners and policy makers, research and extension directors, as well as scholars and practioners who are concerned with agricultural development in poor countries.

This book will also serve as a reference material and highly useful to the community of scholars and scientific workers both in the laboratories and applied fields. This will set the phase for the future generation in this area.

AUTHORS

Contents

Feedback Dharmathupatti

Feed back from the farmers

T8 - Cow dung extract + Zinc sulphate sample of results in Red banana

T10 - Urea + Sulphate of potash + Water + fresh cow dung Sample of results in Red banana

T8 - Cowdung extract + Zinc sulphate Sample of results in Poovan

Feedback Kondayampatti

Advisory committee interacting with farmer

Assessing the growth of the treated bunch

Examining the treated bunch

Explaining the principles involved

Interaction with experimented farmer

Feedback Kondayampatti

Observing feed back

T8 - Cow dung extract + Zinc sulphate Sample of results in Robusta

T9 - Cowdung extract + Borax sample of results in Robusta

T 10 - Urea + Sulphate of Potash + Water + Fresh cowdung Sample of results in Robusta

FS Meet Dharmathupatti

Listening the results of the research

Obataining feed back

Presenting results to the participants

Sharing the results to the farmers

FS meet Kondayampatti

Displaying the content of workshop

Presenting the views on results

Presenting the results to the participants

Sharing experience - Experimenter

Key informants workshop

Briefing objectives of research

Discussion between farmers and Scientists at Dhrmathupatti

Discussion to finalize the treatment combinations at Kondayampatti.

Farmers scientists interaction in Key informant workshop at Dharmathupatti

Finalize treatment combinations in Kondayampatti

Finalizing the treatment combinations

Group leader sharing his experience at Dharmathupatti

Noting down the treatment combinations at Dharmathupatti

Treatments in Kondayampatti

Displaying Treatment details

Enlightening various mixed ingredients by the researcher

Mixing the treatment combinations

T1 - household wood ash + pregnant cow urine

T2 - Brick chamber wood ash + Pregnant cow urine

T3 - Panchakavia + Brick chamber wood ash

T4 - Panchakavia + House hold wood ash

Treatments in Kondayampatti

T5 - Urea

T6 - Urea + Sulphate of Potash

T7 - Cow dung extract + Zinc sulphate + Borax

T8 - Cow dung extract + Zinc sulphate

T9 - Cow dung extract + Borax

T10 - Urea + Sulphate of Potash + Water + fresh cow dung

T11 - Cow dung extract solution

T12 - Control

INTRODUCTION

Tell me and I forget, show me and I may remember,
involve me and I will understand

— Ian Mccelelland, 2000

(PROCEEDINGS OF THE 19th AUSTRALIAN AGRONOMY CONGRESS)

Indigenous Knowledge is the accumulated knowledge, skills and technology of the local people. Indigenous Knowledge has two powerful advantages over scientific knowledge like it has little or no cost, and is readily available. Indigenous Knowledge can be defined as the knowledge built up by a group of people through generations of living in close contact with nature over the years; the farmer in India had been using cow dung and cow urine as manure in the fields.

Indigenous Knowledge is found to be socially desirable, economically affordable and sustainable and involve minimum risk to rural farmers and producers. The failure of modern chemical farming to deliver prosperity to agricultural communities, increase in pest attack of crops, deterioration of soil and water resources, cost to human and animal health has forced scientists to seriously examine whether traditional practices of farmers have many answers to the problems of modern agriculture. Scientists accept that indigenous practices of farmers have pointers for sustainable agriculture in future. Indigenous knowledge systems provide a frame of reference for strengthening agricultural extension programmes and this has led to reorganization of interventions made by extension personnel. Identifying, documenting and incorporating indigenous knowledge in agricultural extension organization is essential to achieve sustainable agricultural development.

Agriculture is not only a means of survival but also a way of life. The application of high input technologies in intensive agriculture has undoubtedly increased agricultural production, but now realizing the plateau in the growth, there is serious concern over their adverse effects on environment quality.

Sustainable agriculture aims at achieving permanence through utilization of renewable resources. Indigenous knowledge is the knowledge acquired by the local people through their past experience, evolved over the time under the influence of traditional knowledge, external factors and agents, and individual innovations. Knowledge, skill and survival strategy of farmers operating with low external inputs have often been ignored to promote modern agriculture. Farmers based indigenous/traditional knowledge has scientific rationale and great deal of relevance for agricultural productivity and sustainability.

Indigenous knowledge or local knowledge is the knowledge that is unique to a given culture or society. It is the basis for decision – making in agriculture at a local level, natural resource management and live stock management. Such knowledge is basically passed down from generation to generation. Indigenous knowledge has the value not only for the culture in which it evolves, but also for scientists and planners striving to improve conditions in farm families and concerned about sustainability in agriculture. The existence and value of farmer's knowledge and innovativeness and experimentation have been widely recognized as a vital input for sustainable agricultural development in recent times. As a consequence to the paradigm shift in technology generation and transfer, efforts are made in various parts of the world to document the farmer's contribution as innovation in the process of technological change.

Over centuries, indigenous people and farmers have developed their own region specific knowledge and practices in agriculture and natural resource management. Farming communities have developed innumerable ways of obtaining food and fiber from animals and plants through a wide range of indigenous farm practices based on generations of experiences, informal experiments and intimate understanding of their biophysical and socio - cultural environments. They had started agriculture as activity very close to the nature and in harmony with all living and non – living things on earth.

India has made tremendous progress and development in agriculture and allied fields especially after the green revolution. As a result, many high yielding varieties, hybrids and frontier technologies have found their way into agriculture. Hence the emergence of technologies and intensive use of inputs without considering their adverse impact on environment and sustainability resulted.

Today attention is shifting to a sustainable form of agriculture to ensure the attainment and continued satisfaction of human needs for the present and more importantly for the future generations. Thus the latest trend world over is unwrapping the indigenous knowledge as an alternative to high input agriculture.

WHAT IS INDIGENOUS TECHNICAL KNOWLEDGE?

Indigenous Technical Knowledge refers to the unique, traditional, local knowledge existing within and developed around the specific conditions of women and men indigenous to a particular geographic area (Grenier, 1998). This complex of knowledge and practices is generally known as 'Traditional Knowledge' (TK), 'Indigenous Knowledge' (IK), and 'Indigenous Technical Knowledge' (ITK), 'Folk Knowledge', Local Knowledge and Wisdom of the elders'. These synonyms refer to the local origin and promotion by a community (Seeland, 2000). Indigenous knowledge is complex and dynamic which are the contributions of traditional wisdom generated through experience, observation, rational thinking, on – farm trials and intimate understanding of the environment in a given culture that enables the communities to survive.

Indigenous practices have two powerful advantages over outside knowledge

It is now widely accepted that in improving traditional farming technologies both the knowledge of the rural people and the knowledge of outsiders are important. The specific way of relating the two sources of knowledge requires a delicate process of formulating, comparing, merging, dialoguing and negotiating between rural communities and outside experts.

Farmer participation in agricultural research is important and necessary first of all to increase the efficiency and impact of agricultural research and technology development. This includes the identification of traits that can guide crop breeders' work. Farmer empowerment is valuable and desirable, and while it can result from participatory research, direct empowerment *per se* should not be the main objective of participatory research conducted by research organizations. Of more importance are the empowerment of partner organizations and the identification of future research needs, i.e. the functional purposes of participatory approaches in agricultural research.

Farmer participation in agricultural research can be defined as a systematic dialogue between farmers and scientists to solve problems related to agriculture and ultimately to increase the impact of agricultural research. While internal rate of return and cost–benefit analyses may have been sufficient for the accountability functions of impact assessment, they do not satisfy those interested in knowing how, and why, a project affects peoples' lives. Impact-assessment practitioners must now document a much broader range of project impacts, especially in the areas of poverty alleviation and environmental sustainability.

In this context, one of the key questions is what are the impacts of participatory agricultural research on rural innovation capacity? Rural innovation can occur in two ways: (i) when external, new technologies become more accessible, e.g. as a result of lower costs of adoption, or (ii) when new technology is developed locally, e.g. due to increase in the local capacity to adopt, adapt or develop new technologies. Different actors can be involved in reducing the costs of adoption of new technologies, e.g. research and extension organizations, private companies, farmer organizations etc. Meanwhile, when rural innovation takes place as a result of an increase in local innovation capacity, the focal point generally is the farmers themselves. This is especially the case when there is an erroneous assumption that the technology is appropriate and desirable, i.e. innovation is seen as a supply issue, when it may be the case the technology is inappropriate and needs adapting so as to complement farmers' needs. The use of participatory approaches is one way of enhancing rural innovation capacity, whether through increased accessibility of externally developed technology or through increased local capacity to address problems and devise solutions for them.

It is now widely accepted that in improving traditional farming technologies both the knowledge of the rural people and the knowledge of outsiders are important. The specific way of relating the two sources of knowledge requires a delicate process of formulating, comparing, merging, dialoguing and negotiating between rural communities and outside experts. Participatory Technology Development (PTD) consists of various phases: getting started, looking for things to try, designing the experiment, execution, sharing the experiences, and sustaining the process. The phase of 'farmers designing the experiment' has been the most problematic. It is relatively easy if the researcher is 'in the driving seat', dictating the pace and direction of the experiments. This position is also reflected in other literature on farmers' experimentation

Participatory Technology Development (PTD) is an approach to learning and innovation that is used in international development as part of projects and programmes relating to sustainable agriculture. The approach involves collaboration between researchers and farmers in the analysis of agricultural problems and testing of alternative farming practices.

Origins of Participatory Technology Development

The origins of Participatory Technology Development can be found in a number of approaches, concepts and techniques that became popular in the 1970s and 1980s:

- Farming Systems Research and Extension (FSR/E), an approach that introduced socio-economic issues into organizations previously dominated by biological and chemical scientists. FSR/E also promoted the idea on 'on-farm' trials.
- Appropriate technology, a concept that recognises the importance of matching the design of technology to the resources and culture of the user.
- Indigenous technical knowledge, a concept that recognises the importance of local expertise and traditional wisdom.
- Participatory Rural Appraisal (PRA), a set of techniques that promotes the involvement of rural people in decision-making.

The PTD process

One of the leading authorities on this process is the Institute for Low External Input Agriculture (ILEIA) based in the Netherlands. ILEIA has described PTD as "a process between local communities and outside facilitators which involves:

- Gaining a joint understanding of the main characteristics and changes of that particular agro-ecological system
- Defining priority problems
- Experimenting locally with a variety of options derived both from indigenous knowledge ... and from formal science, and
- Enhancing farmer's experimental capacities and farmer-to-farmer communication."

PTD facilitators

PTD facilitators are usually researchers, sometimes consisting of a team that includes both biological and social scientists. While PTD is closely linked to research, it often crosses the boundary into

Agricultural Extension because it involves learning activities with farmers.

In the wake of the Farmer First Workshop and Publications in the mid-1980s, participatory approaches to agricultural research and extension were promoted across all continents and groups of development actors. Many of these approaches, together often known under the generic term of Participatory Technology Development (PTD), put small-scale farmers and their concerns at the center of the development agenda. The goal invariably emphasized the strengthening of farmers' capacities to develop, assess and adapt technologies. Joint learning by farmers, extension workers and researchers was the main vehicle for the development of ecologically-oriented agriculture and Natural Resource Management.

By the 1990s, an increasing number of organizations and projects had started implementing PTD in various settings. Often, partnerships were forged between non-governmental and governmental organizations involved in research and development in rural areas. Earlier attempts have captured some of the lessons learnt in field applications, including Joining Farmers' Experiments: Experiences in Participatory Technology Development (1991), Farmer Participatory Research, Rhetoric or Reality (1994), and Farmers' Research in Practice (1997). Those studies were based mainly on evidence going back to the late 1980s and/or focused on a selection of well-known and well-documented pioneering cases.

The field of PTD is wide and the range of experiences diverse. This is largely so because a broader range of actors are involved. Often the nature of the activity is influenced by which agency initiates the activity (e.g. research, extension, civil society institution, etc.) or if it is a multi-stakeholder initiative. There is a need to look beyond off-quoted cases and to learn from the experiences of projects and organizations that have tried to advance the approaches and methods of PTD by incorporating these into larger institutional frameworks. These organizations have encountered many practical constraints in implementing PTD on a larger scale. In doing this, they have modified approaches and have made methodological innovations.

WHY PTD?

In recent years, there has been growing dissatisfaction with the poor rates of adoption of agricultural technologies in resource-poor farming systems. This poor adoption has resulted partly because when agricultural technologies are developed, there is little input from farmers. Participatory Technology Development (PTD) offers a way forward, through active, decision-making involvement of farmers in every stage of technology development.

When new technology is being developed, researchers usually control all steps of the process. There is little, if any, input from farmers. The Participatory Technology Development (PTD) approach involves farmer's right from the beginning, when they are asked to identify their problems.

Problem Diagnosis

The first step is Problem Diagnosis. Researchers work with a representative group of farmers to gain a greater understanding of their agricultural and livelihood systems. The farmers identify the problems that are of most concern to them. They describe what actions they have taken in the past to minimize each problem, and decide which of the problems have the highest priority. They then discuss what action they would like to take to solve these problems in future.

The job of the researcher at this point is to provide farmers with as broad a range as possible of technical solutions and technologies that may help to solve the farmers' problems. Researchers must resist the temptation to offer farmers, only the technology they think is the best. But, farmers frequently use different criteria than researchers when they evaluate agricultural technologies.

Experimentation

Experiments can be formal or informal. Once farmers have developed some promising technologies, controlled experiments can then be conducted to validate and quantify the farmers' experiences.

Evaluation

Once farmers have begun testing technologies and have selected the most promising, there follows a period of evaluation. During this stage, farmers describe which of the technologies they like and why. They also explain which technologies they do not like, and why, and what characteristics of the preferred technologies could be improved.

PTD the Nuts and Bolts

Careful Selection of Sites and Farmers

Careful selection of the sites and farmers is critical. This is often not given the attention it requires. Initial site selection is based on secondary information. At each selected site, researchers should then confirm with farmers that really there is a problem, using problem diagnosis.

Communication

Basic communication skills are the key to successful implementation of PTD approaches. People working with PTD must be able to listen to farmers (not just hear what they say). They must be able to ask probing questions to gain a deeper understanding of farmers' needs, and be able to work in partnership with farmers. These skills are not obtained overnight or from formal training courses, but from field experiences.

Speed of Action

Planning and working with farmers needs to happen rapidly and with commitment. Farmers must not feel that a program is all talk and no action.

Working Partnership

Working in partnership with national government and non-government organizations is essential if we want to foster change in the way agricultural technologies are developed with (and for) smallholder farmers. This requires compromises, commitment and time.

Finding the Right Farmers

Working with a small number of enthusiastic farmers with good rapport and support is more successful than working with many farmers with whom you have little contact.

Minimizing Risks

Smallholder farmers don't maximize their most profitable activities, but minimize their risks.

Farmers' Research and Innovation

Experimentation and innovation are natural and necessary to farmers. Before formal research and extension services existed, farmers' own experimentation allowed adaptation to new situations, to survive and to improve their livelihoods, where conditions were favourable. Worldwide, there are countless cases of farmers adapting extension recommendations to fit particular situations, or extracting components from otherwise seemingly useless packages.

This is still the case today, even where farmers have access to external support. Scientists who develop technology packages for extension seldom realize the extent to which farmers conduct informal experimentation with components of these packages.

Most scientists cannot recognize and understand how farmers experiment the technologies. Yet, many field workers of Non-Government Organizations (NGOs), development projects and extension agencies are appreciating farmers' informal experimentation as a springboard for developing locally appropriate technologies. These development workers and farmers are engaging in PTD, regardless of whether they have support from researchers or not.

Encouraging Farmer Experimentation

PTD where there is no researcher happens wherever there are development workers encouraging farmers to experiment, innovate and adapt new ways of managing agricultural and natural resources. Instead of transferring a "best bet" technology pre-selected by scientists on behalf of farmers, development workers can choose from the following approaches in PTD (Table 1).

Table 1.Participatory Technology Development Approaches to Farmer Experimentation

PTD Approach	Role of Development Agent	Benefit to Farmers
Learning from farmers	• Identifying farmer-innovators • Understanding farmers' experiments • Getting insights into farmers' priorities	• Sharing of experiments among farmers
Testing new options	• Suggesting options and ideas to farmers • Encouraging farmers to compare options and ideas with current practices	• Farmers are free to test, adapt or reject technologies without pressure from development agents
Filling local knowledge gaps	• Enhancing farmers' awareness on resource management principles • Providing information on phenomena that farmers cannot observe on their own	• Local ways of applying principles to farming
Facilitating mutual learning	• Facilitating the generation of insights and options within the community	• Critical exchange of ideas among farmers • Minimal dependence on technology from outsiders

Improved experimental design	• Studying current methods of informal experimentation with farmers • Reaching an agreement with farmers on more systematic forms of experimentation	• Ways of experimenting and learning improved • Position and confidence of development agents and farmers strengthened

Strengths of the PTD Approach

The PTD approach to extension strengthens local research and adaptive capacities by involving a vast number of community-based researchers: men and women farmers and field agents of various government agencies, NGOs and development projects. Research and development planning incorporates local realities, potentials and limitations from the start.

This approach greatly reduces the time spent between problem identification and development of applicable solutions, especially for problems that can and should be tackled at the farmers' level. Results from site-specific, farmer-led research and innovation in one locality can rarely be replicated exactly elsewhere but can serve as sources of ideas for farmers in other areas.

PTD where there is no researcher is a cost-effective approach as it does not require highly-paid scientists. Development agents involved in the approach live closer to the farmers, use local services and facilities, and tailor their work to use resources at hand more efficiently. Focusing on local knowledge and resources makes development agents and farmers more equal as partners in their research pursuits.

Challenges

Many development agents lack the confidence to enter into this open-ended approach because of possible sanctions for not meeting expectations in transferring technologies from research stations. Development agents are restricted in encouraging farmers to try things on their own terms on a small scale. Situations must be created that lead to attitudinal change of development agents and their superiors towards accepting that farmers' knowledge and innovation are complementary to their knowledge and skills.

Substantial time is needed to support PTD training, with brief learning sessions being interspersed with longer implementation periods. Learning sessions should include real cases of farmers- or community-led experimentation, and design of follow-up assignments by the trainees themselves. At the end of each implementation period, trainees should reflect jointly on their experiences.

Public funds for extension services to farmers and the number of development agents are both decreasing rapidly as privatization expands. The same applies to research, where scientists are obliged to do work more relevant for farmers. This can be done by linking with farmers and development agents in PTD where there is no researcher. Scientists and policymakers seeking such links have to re-think their definition of research and its interface with development.

Participatory and other approaches to technology development have shared a recent preoccupation with specific methods and doubts about just how much can be expected of the methods themselves, as opposed to how they are applied, by whom, and in what circumstances. Detailed analysis of historical cases suggests that the development of both technologies and methodologies is highly dependent on local context. Such processes are characterized by conflicts over the direction of change and affected by the activities of a particular type of grouping, the development coalition. These coalitions are examined and implications are considered for training, education and Participatory technology development.

The success of PTD where there is no researcher cannot be judged by conventional scientific criteria, but by contribution to solving local problems.

PTD Interactions through Peer Exchange and Feedback Mechanisms

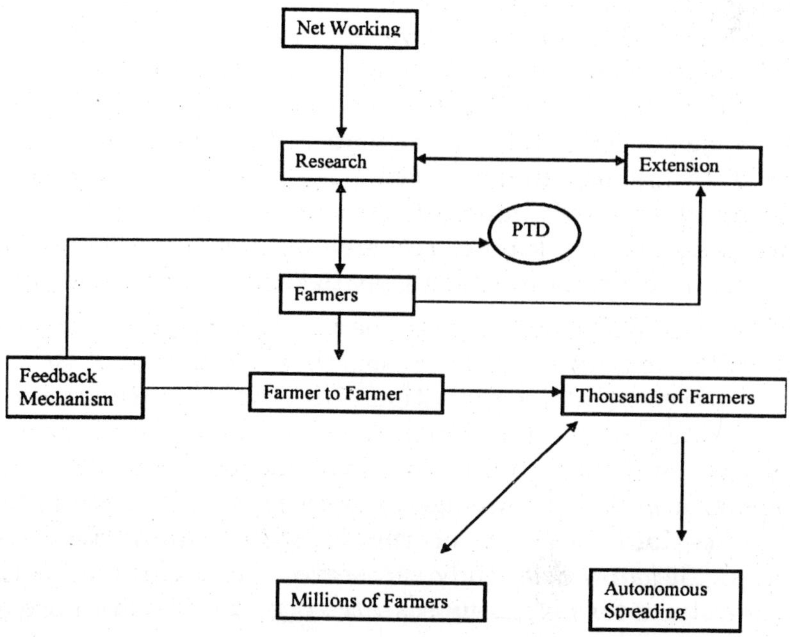

Keeping this in view, a research was conducted with the following objectives.

Specific Objectives

1. To study the Socio – Personal and Psychological characteristics of the farmers in the study area.
2. To identify the Indigenous practices amenable for Participatory Technology Development (PTD).
3. To operationalize the steps involved in PTD Process and build the platform for Joint experimentation for farmers and Researchers.
4. To refine the identified Indigenous Agricultural Technology through Experiments for better adoption
5. To compare the refined Indigenous Agricultural Technology with the scientific one.
6. To assess the usefulness of the technologies developed and the scope for up scaling the PTD process.
7. To test the refined technologies for successful advocation.

Scope of the Research

The concept is a pioneering one in Tamil Nadu and a Novel attempt in the field of agricultural Extension. Once the methodology is standardized, the protocol can be used for validation in other crops and enterprises with extrapolative effect. It will help to streamline programmes for increasing yield and productivity in Banana crop and other crops of farm land. The Participatory Technology Development provides a model for involving farmers in technology generation and transfer of technology under varying agro-socio-economic conditions. This study will have immense practical utility in the effective implementation of this kind of research on a wider basis for different agricultural enterprises in varying spatial and temporal situations.

The rationalization analyses may give confidence to the client system, extension system and research system either to accept or reject the treatment combinations in the identified Indigenous Technical Knowledge in Banana. The evaluative perception of the respondents on the treatment combinations in the identified Indigenous Technical Knowledge in Banana would throw light on the practicability and viability of the identified indigenous technical practices in Banana. The study can serve as a useful feed back to the research systems for designing research agenda, research projects and on farm trials for testing, validation, refinement and blending of various Indigenous Technical Knowledge with modern technologies for large scale recommendations.

The study is being a Participatory Technology Development (PTD), there is an ample scope for the farming community to involve in each and every step followed in the technology generation as and when required under different agro – ecological situations, agro – economic and specific micro farming situations.

Limitations of the Study

The present research formed as a part of the Doctoral degree programme investigation and hence it has all the limitations of time, money and other resources. These limitations determined the restricted selection of localities, farmers and sample size for the study. However, careful and rigorous procedures were adopted to carry out the research systematically.

Indigenous knowledge systems being highly location specific in terms of specific needs and resource availability and the results of the study may not be generalized to all the farming communities since specific technologies will have to be designed depending upon the changing micro situations.

The success of Participatory Technology Development (PTD) depends on the active co-operation of various sub systems, agencies and participant farmers and obviously control of the researcher over the experimental programme also inevitable.

In spite of these, it is believed that the findings of the study and conclusions drawn would stand the test of more rigorous field situations. It is expected that the refinement of identified indigenous technical knowledge in Banana through Participatory Technology Development would be an indispensable record for the present and future generation of farmers, extensionists and scientists.

THEORETICAL ORIENTATION

A review of previous research studies help to identify new problem areas and research priorities and provides a basis for developing a theoretical frame work and methodology for the study. In accordance with the specific objectives set, the review of literature is furnished below under the following sub heads:

2.1. Indigenous Technical Knowledge (ITK) / Indigenous Knowledge (IK) – definition and concept

2.2. Nature and importance of Indigenous Knowledge

2.3. Methods of collection, documentation and Experimentations of Indigenous Knowledge/practices

2.4. Perceived effect and measurement attributes of Indigenous Knowledge

2.5. Rationality of Indigenous Knowledge

2.6. Integration of scientific and indigenous wisdom

2.7. Concept and advantages of Participatory Research and Participatory Technology Development (PTD)

2.8. Traditional technologies used for agricultural purposes

2.1. INDIGENOUS TECHNICAL KNOWLEDGE (ITK) / INDIGENOUS KNOWLEDGE (IK) - DEFINITION AND CONCEPT

Past researches conducted and articles published in this field had defined and emphasized on the concept of indigenous knowledge/ practice in different dimensions and they have been detailed below:

Indigenous knowledge is defined as a system fine tuned and adapted both biologically and socially to counter the process of what are often harsh and inimical environments and often represents hundreds or thousands of years of adaptive evolution in which the vagaries of climate, the availability of land and water, the basic needs of people and their animals for food, shelter and health have been amalgamated in a system which has helped societies to exist and develop in the face of tremendous odds.

Warren (1987) defined indigenous knowledge as local knowledge, knowledge that is unique to a given culture or society. This knowledge is the information base for the society, which facilitates communication and decision making. Indigenous knowledge is passed down from generation to generation by word of mouth.

Carter (1988) while describing the nature of indigenous knowledge indicated that IK is highly localised and restricted. Local environmental factors and cultural conditions govern the revolution of IK these vary between countries, regions and even farm to farm.

Indigenous knowledge (IK) is the sum total of knowledge and practices that are based on peoples accumulated experience in dealing with situations and problems in various aspects of life and such knowledge and practices are special for a particular culture (Wang, 1988). He also defined IK as the knowledge inbuilt in the system itself by virtue of informal documentation (or) practices evolved by the farmer clients over the years to generate technologies that have been proven sustainable.

Warren & Cashman (1988) explained local knowledge as the sum of experience and knowledge within a given group, which forms the basis for decision – making, related to familiar and unfamiliar problems and challenges.

Warren (1989) identified indigenous knowledge as it's unique to a given culture or society. This knowledge is the information base for a society. It facilitates communication and decision – making. It is dynamic and it changes through knowledge creativity and innovativeness as well as through contact with other knowledge system. He also explained that the information gained over a period was passed on from generation to generation by word of mouth. This knowledge in today's parlance is called local knowledge, traditional knowledge or simply IK.

Chittiraichelvan and Raman (1990) explained that IK includes both technical and non-technical fields covering various social and religious taboos, beliefs and customs, communication patterns, music, ecology vegetation, climate and so on.

Indigenous Knowledge, as stated by Warren (1990), is the systematic body of knowledge acquired by local people through the accumulation of experience, informal experiments and intimate understanding of the environment in a given culture.

According to Altieri (1991) local knowledge can be defined as the accumulated knowledge, skills and technology of the local people, derived from the direct interaction of human beings and their environment.

Goldman (1991) stated indigenous knowledge as reflecting climatic and socio-economic factors, embedded as they are in social organization, cultural traditions and preferences and even more fundamentally in the conceptual system in which the individual members of the society have learned to think and in terms of which they interpret their society and environment.

According to Haverkort (1991) opined that the knowledge of a given population that reflects the experiences based on traditions and includes more recent experiences with modern technologies. Local people, including farmers, landless laborers, women, rural artisans and cattle rearers, are the custodians of IK systems.

According to Rajulasanthy (1991) Indigenous resource management is the art or manner by which the available local resources are directed in a better way to improve the income.

Indigenous knowledge is the local knowledge – knowledge that is unique to a given culture or society (Warren, 1991). It is the basis for local – level decision making in agriculture, health care, food preparation, education, natural resource management and a host of other activities in rural communities.

Wilson (1991) revealed that local knowledge system are strategic resources for decision – making in societies seeking to maintain their priorities with respect to sustainable production and to preserve those elements of society which are essential for social and demographic stability.

Indigenous knowledge is the knowledge of the people living in a certain area, generated by their own and their ancestors' experience including the knowledge that originated from elsewhere and has been internalized by the local people (Reijntjes, 1992).

Reijntjes *et al.* (1992) opined that the knowledge of a farming population living in a specified area is devise from the local people's past experience, both that are handed down from previous generation and that of the present generation.

Talawar and Singh (1992) stated that IK has undergone evolutionary process and is built from and based on thousands of years of experience

Warren (1992) clarified that the term Indigenous Knowledge (IK) is used synonymously with 'traditional' and 'local' knowledge to differentiate the knowledge developed by a given community from the international knowledge system sometimes also called 'Western' system, generated through universities, government research centres and private industry. IK refers to the knowledge of indigenous people as well as any other defined community.

Berkes *et al.* (1993) reported that traditional people who depend on their local ecosystems for their essential needs, have accumulated by trail and error a rich body of local environmental knowledge and in several cases elaborated resource management systems and development institutions appropriate for implementing these systems. Biodiversity conservation appears to be the most integral part of much traditional management system from tropical forests to coastal fisheries.

Browers (1993) defined the terms 'indigenous' and 'local knowledge' are used to refer to that knowledge which is generated and transmitted by communities, over time, in an effort to cope with their own agro ecological and socio–economic environment. The knowledge is generated and transformed through a systematic process of observing conditions, experimenting with solutions and readapting previously identified solutions to modified environmental, socio–economic and technological solutions.

Ulluwishwa (1993), IK is a people – derived science and it represents people's creativity, innovation and skills.

Fernandez (1994), indicated that the terms 'indigenous" and 'local' knowledge are used to refer to that knowledge which is generated and transmitted by communities, over time, in an effort to cope with their own agro ecological and socio – economic environments.

Rhodes and Bebbington (1994) stated that farmers using IK system engage in systematic experimentation either for problems solving or out of curiosity create new technologies and test or adopt existing techniques and technologies.

Agricultural practices that are evolved locally and inherited over a long period of time are referred to as indigenous practices (Talwar and Singh, 1994).

Titiola (1994) stated that local knowledge is an important aspect of society's cultures. Traditional knowledge, which is based on accumulated experience, much fitted to the local situation and social system. It is dynamic in nature.

According to Vasu (1994) indigenous know – how has two connotations. One is concerned with the traditional technologies and the other is with respect to the modern technologies, either developed indigenously or imported and adopted to indigenous conditions.

Agarwal (1995) discussed that because IK changes overtime, it is sometimes difficult to decide whether a technology or practice indeed is indigenous, or adopted from outside, or a blend of local and introduced components. In reality, there is a lot of overlap between indigenous and western knowledge, and it can be very difficult to distinguish between them.

Parvathi (1995) defined IK as the knowledge in – built in the system itself by virtue of informal documentation or practices evolved by the farmer clients over the years to generate technologies that have been proven sustainable. It was also defined as the sum total of knowledge and practices, which based on people's accumulated experience in dealing with situations and problems in various aspects of life and such knowledge and practices are special to a particular culture.

Somasundaram (1995) used the term "IK" synonymously with 'traditional' and 'local knowledge' to differentiate the knowledge developed by given community from the international knowledge system sometimes also called as 'western system' generated through universities, government research centers and private industries. He further stated that Indigenous Knowledge is traditional knowledge that is based on accumulated experience, much fitted to the local situation and social system. It is dynamic in nature.

Alamelu (1996) stated that the information developed by farmers over a period of time was passed on from generation to generation by oral words of mouth. This knowledge is definitely called as ITK.

International Institute for Rural Reconstruction (IIRR) (1996) stated that the following are the characteristics of IK (or) local system viz., most local people are generalized, IK system are holistic, IK systems integrate culture and religion and IK systems minimize risk rather than maximize profit.

In the terms of Lawas and Luning (1996) Indigenous knowledge denotes a type of knowledge that has evolved within the community and has been passed on from one generation to another.

Preetha (1997) has operationally defined indigenous practices as resource saving, site specific, farmer devised technologies experimented and adapted by themselves which is simple to practice, flexible in use, and sustainable in effect.

Rambabu (1997) defined the indigenous technologies as the technologies developed by the local farmers based on the topography, local agro climatic conditions and available resources through non – formal experiments.

Brouwer (1998) pointed out that indigenous knowledge, in general is the participant's knowledge of their temporal and social space. For purposes of clarity, Indigenous Knowledge must be distinguished from traditional knowledge. True tradition comprises proven, ancient, original and distinctive customs, conventions and routines. Thus, tradition operates on the practical level of repeated actions based on opinion or belief. The actors need not have any knowledge indigenous or otherwise, to successfully carry out and pass on their traditions.

Indigenous knowledge refers to unique, traditional, local knowledge existing within and developed around the specific conditions of women and men indigenous to a particular geographical area (Grenier, 1998).

As defined by Marrewijk (1998) indigenous knowledge is the sum total of the knowledge and skills that people in a particular geographic area possess and which enable them to get the most out of their natural environment. Most of this knowledge and these skills have been passed down from earlier generations, but individual men and women in each new generation adapt and add to this body of knowledge in a constant adjustment to changing circumstances and environmental conditions.

Berkes (1999) has defined the Indigenous Knowledge as the local knowledge held by indigenous people, or local knowledge unique to a given culture or society.

The knowledge derived through trial and error with many crops and practices and with sharing of knowledge within many farming families: which are crop, climate and soil specific are referred to as indigenous knowledge (Babu, 2000).

According to Eyzaguirre (2001), indigenous knowledge in the form of local know – how and cultural practices is a set of tools that communities used to manage their natural resources, which include genetic resources, the building blocks of bio diversity and agriculture.

Seetharaman (2001) stated that our ancestors have developed, modified, defined and perfected the knowledge developed by them. Such knowledge item has been forwarded to this progenies, thus came to existence a fund of knowledge called 'traditional knowledge' or IK.

The information gained over a period of time was passed on from generation to generation by word of mouth. This knowledge in today's parlance is called local knowledge, traditional knowledge or simply indigenous knowledge (Gupta, 2002).

To summarise, ITK/IK in agriculture are used synonymously to indicate farmers' practical knowledge about their local production system, their farming techniques and skills to manage with their natural resources to gain the basic needs with sustainability. It is dynamic and unique to a given culture or society.

2.2. NATURE AND IMPORTANCE OF INDIGENOUS KNOWLEDGE

Earlier, studies on indigenous knowledge mostly focus upon agriculture, agro – forestry and livestock farming. Only recently, there have been some attempts to study the role of indigenous knowledge in rural development and natural resource management.

The study of indigenous knowledge is important in planning and implementing new progarammes. These are the entry points for future scientific work and development of appropriate and acceptable practice (Chitamber, 1961; Sikhana, 1994).

Biggs and Tinnermeir (1974) narrated that there is much to be learned from traditional resource management practices being utilized on small farm through out the developing world.

Andrew (1978) noted that in Japan the level of agricultural Technology has been raised steadily since the end of last century by means of indigenous research on the basis of existing plant resources and systems of production.

Garg and Hoda (1978) stated that maximization of production influences the income of the farmer to a substantial extent, but if the inputs for achieving high production are provided from outside the rural society, then a large part of the agricultural income will be drained to the urban areas. It is therefore necessary that the technology by which the various inputs are intended to be provided, should be carefully considered as to whether they come from rural society or from outside.

Gladwin (1980) provided a case in which prior ethno – scientific research would have enabled agricultural research to be more responsive to local conditions.

Srivastava (1980) opined that since traditional technologies have undergone a selective process over centuries of empirical testing they are very likely to represent optimal solutions. He also quoted the traditional tools and techniques should be studied and organized effort made by improving their efficiency and productivity.

Alcorn (1984) and Hunn (1985) opined that farmers' practical knowledge about the local ecosystem is reflected in their farming techniques and in their skill in using the natural resources to gain their livelihood. They may reveal ideas, which contain 'seeds' of adaptive value.

Altieri and Anderson (1986) suggested that incorporating indigenous knowledge in the technology development for the resource poor farmers will help in effecting moderate to high levels of food production.

Dommer (1988) observed that a growing number of individuals and organizations have come to appreciate the importance of working with and through indigenous knowledge systems in promoting sustainable agriculture instead of trying to work around them.

Reddy and Narayana (1988) pointed out that the modern agricultural is like a cracked earthen pot, which cannot be used but thrown. He suggested that the agricultural technologists should get out of their ivory towers. Understand farmer's problems and the situations and find out solution. There is need for new technologies adopted by successful farmers to be incorporated in the package of practices.

Compton (1989) opined that incorporating indigenous knowledge in technology development would undoubtedly increase institutional support to sustainable agriculture increase the academic interest and incorporation of indigenous knowledge in the curricula of universities etc., in formulating a suitable paradigm for development.

Scoones (1989) pointed out that understanding farmers' knowledge allows a frame work of reference for posing technical and scientific questions in research. It also provides the basis for evolving technologies, options that are not imposed as alien "packages" which contradict existing practices.

Chittiraichelvan and Raman (1990) said that the indigenous knowledge is the product of centuries of trial and error, natural selection and keen observations that can form a knowledge base on which researchers and extension workers can plan their research strategy and experimental procedures.

Gupta (1990), Reijntijes *et al.* (1992) and Altieri (1996) stressed the need to revise the ITK of farmers from the threatening impact of green revolution and associated attempts to boost agricultural production.

Thurstone (1990) indicated that traditional agricultural practices must be understood and conserved before they are lost with the rapid advance of modern agriculture in developing countries.

Gill (1991) concluded that remarkable degree of institutional and organizational development lies behind the success of the indigenous systems. He added that the so called "uneducated" farmers are capable of creating and maintaining large and complex multimember systems to achieve mutually beneficial results.

Kshirsagar (1991) pointed out that traditional strategies continue to offer important pointers for future adoptive research that aims to find solutions, which will be readily acceptable to farmers.

Pretty *et al.* (1991) reported that Indigenous Knowledge Systems are tuned to the needs of local people and the quality and quantity of available resources.

Sanghi and Kerr (1991) in their paper on soil and water conservation practices concluded that researchers and extensionists have much to learn from farmers regarding cost effective and relevant methods of soil and water conservation.

Indigenous knowledge is mainly inherited through the socio cultural system, which is maintained and developed through oral traditions, folk tales and proverbs (Verma and Dhukia, 1991; Atte, 1989; Vijayalaxmi, 1996).

Warren (1991) revealed that there is a growing body of evidence describing the positive users of Indigenous knowledge in development projects and many successes in building upon it. Indigenous societies possess complex and sophisticated knowledge about their environments and natural resources which is dynamic and constantly changing as people try to adapt to new conditions and seeks solutions through experimentation and innovation.

Balasubramanian (1992) opined that farmer initiated technology does not occur by accident, but there is a farmer based method of research, similar to scientific method. It is concrete and relies strongly on institution, historical experience and directly perceivable evidence.

As observed by Drinkwater (1992), indigenous knowledge is left untapped by development agencies despite the fact that farmers continue to use, perpetuate and build on this type of knowledge. It is generated and exchanged with in the informal knowledge networks.

Moockes and Rhoades (1992) argued that along with biological science, past and present indigenous farmer knowledge can play a key role in sustainable agricultural development.

Roling and Engel (1992) warned that a look at farmers only as users neglects the important fact that farmers are experimenter and that farmers have developed most of the technology used on the farm today.

Talawar and Singh (1992) state that for some, IK provide a basis for identifying ecologically sustainable options of resource use and for others these are cheap sources of identifying ideas, which have considerable scope for commercial exploitation after value addition.

Bently (1993) suggested that scientists could learn more from farmers about factors in the farmers' lives that are important and easy to observe.

Berkes *et al.* (1993) argued that in order to ensure a more socially and ecologically sound approach to development, it was necessary to understand, respect and utilize the local knowledge systems.

Bharara and Sceland (1993) emphasized the importance of the consideration of IK in the development process of areas with precarious environmental conditions in their contribution, they highlighted the immense value of autochthonomous perceptions of particularly rural surroundings to take future challenges and to safeguard the sustainable, physical, social and cultural persistence of the semi – arid area of Rajasthan.

Kanagasapapathi (1993) reported that those indigenous practices play a vital role in sustainable agriculture. He added that certain practices shatter the belief that recommendations made by agricultural scientists are always superior.

Pereira and Gupta (1993) pointed out that one reason for the persistence of traditional technologies and their revival now is that westernized system cannot solve the problems that individual farmers and artisans face either because of the seed for urgent action, or because westernized solutions are too expensive or because they are seen to damage the environment.

Rajasekaran and Whiteford (1993) indicated that in all agrarian societies, indigenous food production systems form the basis of food and nutritional security. They insisted that indigenous knowledge can be used to fulfill socio – economic needs and conserve biodiversity at one and the same line.

Rajasekaran *et al.* (1993) stressed that identifying, documenting and incorporating indigenous knowledge systems into agricultural extension organizations is essential in order to achieve sustainable agricultural development.

Vijayalaxmi (1993) suggested that knowing about and enriching tradition in indigenous agricultural practices will help in finding ways of continuing with high yield farming without poisoning the soil, water and air with pesticides and fertilizers.

Ahmed (1994) pointed out the need for a blend between IK and modern technology.

Pamela (1994) pointed out that the indigenous knowledge systems have a great deal to offer in terms of genetic resources, food, medicines, clothing, shelter, fuel, tools, techniques and crop and animal protection. These systems are varied, adaptable, nature friendly and produce yields that are not necessarily lower than those of modern agriculture and concluded that in the absence of appropriate modern alternatives, indigenous knowledge has become a starting point for academic and other institutions in their search for solutions.

Rajapakse (1994) observed that the promotion of traditional food varieties that require minimum inputs under favourable soil and weather conditions can supply a stable yield as an alternative policy scenario to improve the food security of the rural household.

Salas (1994) pointed that one of the main reasons why conventional developmental approaches had failed was that they had tended to ignore the local knowledge systems and practices.

As stated by Titilola (1994) local knowledge is an important aspect of society's culture. He also added that knowledge of local techniques can help to identify practices suitable for adoption of or adaptation, with a view to improve or reinforce accepted methods and procedures without destroying local environments and societies.

Indigenous information systems are dynamic and are continually influenced by internal creativity and experimentation as well as by contact with external systems (Flavier, 1995; Paul and Ramanathan, 2002).

Lupanga *et al.* (1995) conducted a study, which, examined ways of linking research, extension and farmers through a two – way exchange of both indigenous technical knowledge (ITK) and scientific knowledge in order to increase food production in Tanzania.

Vel (1995) pointed out that farmer's indigenous technical knowledge plays an important role in deciding about agricultural innovations.

Altieri (1996) stated that, in this new emerging conception of agricultural development, rural people's knowledge about plants, soils and animals gains unprecedented significance. He continued that scientists involved in small farm development must quickly systematize and incorporate farmer's knowledge before the wealth of practical knowledge is lost forever.

International Institute for Rural Reconstruction (IIRR) (1996) stated the following reasons for IK usefulness i.e. IK is the basis for self – determination for at least two reasons viz, people are familiar with indigenous practices and technologies, it draws on local resources, IK provides effective alternatives to western know – how and IT practices are often cheaper than western ones and further stated that IK systems are holistic in nature; they integrate culture and religion and minimize risk rather than maximizing profit.

Kanagasabapathi (1996) reported that indigenous agricultural practices are cost – effective, time – tested, eco-friendly and they serve to sustain the agricultural development.

Khanna and Bisa (1997) opined that indigenous knowledge is based on traditional wisdom and ethno science, which has been evolved over generation as a product of man and environment interaction. It may be pointed out that ITK is dynamic, changing and at times borrows from other knowledge systems through contact and by gathering first hand experience.

The basic component of any country's knowledge system is IK (World Bank, 1997). It encompasses the skills, experiences and insights of people applied to maintain or improve their livelihood.

In the terms of Emadi (1998) the context of indigenous knowledge is the local environment, in all its cultural, social, economic and physical aspects. To meet demands of the ever – changing nature of local conditions, people have had to be creative and resourceful.

Kashem (1999) stated that the use of IT/ITK by rural men and women farmers, often with slight modifications and or reinvention may be very important in respect of sustainable agricultural development at the one hand and to save the environment on the other hand.

Mendoza (1999) recorded that when faced with difficulties in implementing their activities, farmers try to innovate or adopt existing innovations, depending on the available resources.

The studies made by Sulaja (1999); Mandal and Chauhan (2001) on endangered skills revealed that identification, validation and re appropriation of relevant indigenous farm technologies form the starting point of on farm research. Such efforts can form the base for sustainable and eco-friendly agricultural development.

Gavruek (2000) stressed the need for respect for the cultural life styles of indigenous peoples, because traditional knowledge can build on the scientific knowledge of our country's aquatic ecosystem.

As propounded by Gupta (2000), local knowledge has a tremendous potential for restoring the economic and ecological balance.

Even though traditional knowledge is declining at an accelerated rate, this knowledge coupled with biological diversity is essential to maintain the options for the survival of mankind in a changing world (Juma, 2000; Klemn, 2000).

Neto (2000) opined that indigenous techniques and knowledge are significant in areas such as environmental assessment, resource management and sustainable development.

Parvathi *et al.* (2000) recorded that indigenous knowledge in post – harvest operations is highly valued, since in many cases the new tools and techniques are not available or are beyond the means of farmers.

Sound traditional agricultural practices need to be studied critically and evaluated under farmers, resource management for inculcation of improved components of technology under existing weather and micro farming situations for higher productivity and sustainability (Singh *et al.,* 2001).

Benz *et al.* (2002) opined that traditional knowledge may be able to survive the modernization process today where such knowledge has an important role in subsistence.

It is of utmost importance that farmers' knowledge is taken into account before any agricultural technology is developed and disseminated to its users (Kieft, 2002).

Indigenous knowledge is held in the rural society, usually based on the experience of many generations and unique to each cultural group (Rani *et al.,* 2002). Usually it contains more information on local diversity and complexity that scientifically derived knowledge, which could serve as the basis for sustainable agriculture.

The nature of indigenous knowledge is highly localized and restricted. These technologies have been developed by the local people, refined through their wisdom and is practiced over the centuries (Samantha, 2002). These vary among countries, regions and even from farm to farm.

Sundaramari *et al.* (2004) pointed out that Indigenous technologies are considered effective as they have been tested by the users themselves over a long period of time and perpetuated.

Baldeo Singh *et al.* (2007) stated that the time – tested principles and practices of Indigenous Technical Knowledge (ITK) have amply demonstrated that much of the problems related to natural resource management and ecology degrading agricultural practices could be effectively managed by their utilization either in their present form or improvisation and even blending with modern technology.

From the above review it could be concluded that the indigenous practices are important in the process of agricultural development. This knowledge provides a basis for identifying ecologically sustainable options of research use that are time tuned, both biologically and socially.

2.3. METHODS OF COLLECTION, DOCUMENTATION AND EXPERIMENTATIONS OF INDIGENOUS KNOWLEDGE/ PRACTICES

Various sources and methods have been suggested by different authors for effective collection and documentation of farmers' indigenous practices.

Gupta (1990) listed the following reasons for documentation and experimentation of indigenous knowledge/practices:

- To understand scientific rationale
- To accelerate technological change
- To evolve better understanding of technology development and development of newer concepts
- To increase awareness among the younger generations and develop appreciation for the traditional systems
- To revive and restore pride among the farmers themselves.

Thurston (1990) indicated that traditional agricultural practices must be understood and conserved before they are lost with the rapid advance of modern agriculture in developing countries.

Massaquoi (1993) reported that the method of obtaining information on the indigenous technology would depend on the particular technology and could include two or more of the following:

- Observation of the technology in operation
- Interviews with users and

- Controlled trials of the traditional technology

Rajasekaran (1993) recorded indigenous knowledge systems using farmer participatory methods such as participant observations and unstructured exchanges.

Rath (1993) was of the view that participatory approach was the method for indigenous research.

Vivekanandan (1993) mentioned that conducting village level workshops, group discussion with farmers, artisans, publication of newsletters in local language for exclusive communication of traditional farm technologies and traveling to interior regions were some of the effective means in identification, documentation and dissemination of traditional technology and also a means of getting feedback from the people.

Arulraj and Vasanthakumar (1996) utilized Participatory Technology Development (PTD) approach to identify indigenous knowledge and to integrate the knowledge in technology development process.

As reported by Hanyani and Hebinck (1996), the important sources of indigenous knowledge are innovative farmers, indigenous experts, opinion leaders and village elders.

2.4. PERCEIVED EFFECT AND MEASUREMENT ATTRIBUTES OF INDIGENOUS KNOWLEDGE

Chakravarthy (1982) found that indigenous farm practices in general were perceived to be more culturally compatible, safe, physically compatible, simple and flexible.

Chittiraichelvan and Raman (1991) and Rajaram *et al.* (1991) stated that despite advances in dry farming research, the rain fed farmers depend more on traditional practices involving less cost, having ecological and farming system adaptability and providing more or less stable productivity under aberrant weather conditions to contain the risk.

Socio economic factors affect the farmers' perception towards indigenous knowledge (Kumar, 1994). It was found that acceptance of innovation was dependent on cost effectiveness in the case of big farmers and the compatibility with established procedures in the case of small farmers.

The necessity of labour evaluative perception of ITK items by farmers were highlighted by Vasu, 1994.

Babu (1995) reported that the indigenous practices commonly adopted by the homestead farmers were found to be highly cost effective, less expensive, and easy to practice.

Indigenous knowledge systems and technologies were found to be readily available, socially desirable, economically affordable and sustainable and involve minimum risk to rural farmers and producers and above all, they are widely believed to conserve resources (Grenier, 1998).

The studies conducted on traditional beliefs of farmers commented that the scientists have turned their eyes into sustainable development. Those beliefs that are considered as primitive leftovers of the past have gained much significance. These age old time tested beliefs are not only environmental friendly but also cost effective (Sunil, 1998; Sulaja, 1999).

Ravi kumar *et al.* (2002) Suresh and Hegde (2002) identified certain advantages of indigenous technical knowledge like cost effectiveness, location specificity, suitability with farmers' needs and situations.

In general, indigenous practices were perceived to be more culturally compatible, safe, simple, flexible and sustainable. However, the level of perception varies among farmers, extensionists and scientists.

2.5. RATIONALITY OF INDIGENOUS KNOWLEDGE

Any practice, considered as valid and fruitful have its scientific background for the successful results. Though the farmers are not able to explain such scientific rationale behind those indigenous practices, scientists are responsible for test verifying those practices and finding out their rationality. Studies reporting on the rationality of indigenous practices have been reviewed and presented below:

A survey conducted by Gupta (1987) concluded that scientists perceived peasant farming practices as intriguing, sub optimal and unscientific.

Talwar and Singh (1992) unraveled the rationality of some of the seed techniques in arid regions of Karnataka state. The indigenous practices were based on and confirmed with the logics envisaged in the modern sciences. In this direction, the aim of probing with scientific rationality of IKS was achieved by abstraction and conceptualization of IK and further by comparing with the findings of the formal research and development system.

Pulmate and Babu (1993) studied the scientific rationale of certain existing traditional practices in shifting cultivation in Manipur state. They concluded two–thirds of practices were scientifically rational.

Ramaraju (1993) reported that the practices, scientists considered as more rational were selection of big sized ear heads of millets for seed purpose, erecting the bamboo stick tied with bird feathers to avoid bird attack, keeping maize cobs without removing husk and tussle for seed purpose, red gram seeds were mixed with ash and covered with red earth, paddy seed stored in a ball shaped puri, pulse seeds were mixed with ash for storage, practices of stone bench, terracing in steep hill slopes, practices of stone bunding across the slope, earthen check dam constructed for storage of steam water, short duration paddy varieties suitable to rainfed area were used in a mixed cropping, the dropped leaves provide nutrients to the crops, pineapple grown under shade in banana, mango and cashew orchards, rationing of pineapple after eighteen months and one to two suckers are allowed to grow and others removed in a ratoon crop.

Talawar and Singh (1993) found that the scientific rationale of indigenous practices should be tested based on knowing the context of practice and comparing with the findings of formal agricultural science.

Gupta (1994) stated that local solutions developed by farmers themselves need to be identified and their scientific bases understood. The value – added scientific principles have to be shared back with farmers, who would then be able to develop technologies through their own research and experimentation. Thus transferring "Science" and not just technology.

Babu (1995) and Kurup (2000) stressed that the scientific rationale of many of the indigenous practices has to be looked into systematically which could benefit mankind in many ways.

As reported by Somasundaram (1995) in his study, out of the total Indigenous Knowledge Items (IKIs) related to crops and general agriculture, 59.87% of them were judged as rational and 40.13% were judged as irrational IKIs. More than 70% of IKIs on red gram, pearl millet and sugarcane, more than 60% in the case of sorghum, black gram, ground nut and general agriculture and more than 50% in rice were rated as rational. Rational items were less in the case of cotton, finger millet and gingelly and their percentages being 47.62, 25 and 25 respectively. Under general agriculture, all items related to preparatory cultivation, sowing and economics were judged as rational. In manures and manuring, the rational items were 83.33%

followed by 75% in storage pest management, 71.43% in weed management and 68.42% in soil and water management. More number of items related to crop production aspects on rice, sorghum, black gram, groundnut and general agriculture were found to be rational.

Prasad et al. (1996) in their study on 'rationale of indigenous post harvest practices in Ranchi District' concluded that scientists favoured the continuation of nine out of 11 indigenous post harvest practices on threshing, dehusking and storage activities for rice, wheat and grain legumes by a sample of 200 farmers.

As reported by Rambabu (1997) in his study, out of 340 indigenous practices taken for further study, 267 practices (78.53%) were considered as rational while 73 practices (21.47%) were judged as irrational.

Most of the traditional practices lack scientific basis, thus there is an urgent need to validate the local knowledge (Preetha, 1997 and Ahamad *et al.*, 2002).

A study on traditional veterinary practices among the people of northern plains of Uttar Pradesh by Tripathi *et al.* (1997) established that their beliefs were scientific.

There must be a deliberate attempt on the part of educational institution to find out the scientific relevance of the traditional practices for continuous use (Verma *et al.*, 1997).

Ganeshmoorthi (2000) revealed that 100% of the practices related to preparation before threshing, cleaning winnowing, processing and packaging for storage were rated as rational.

According to Lakshmanan (2000) out of the seven indigenous farm practices in groundnut, 6 were rated as rational and the remaining one is rated as irrational by the scientists.

Sundaramari (2003) reported that 85.49% of 386 indigenous agricultural practices were judged as rational while 14.51% were rated as irrational by the scientists.

Swapna *et al.* 2005 reported that Indigenous practices are based on and conform to the logics envisaged in the modern science. In this direction the aim of probing into the perceived effect and scientific rationality of traditional knowledge system was achieved with the help of the major participants in the field of agriculture. Considering its perceived effect together this can form a base for a successful Participatory Technology development with the help of Extensionists and traditional researchers in the field of agriculture.

From the above reviewed literature it is clear that there must be a deliberate attempt to find out the scientific rationality behind each ITK item/ Indigenous knowledge, besides its mere documentation. The above mentioned studies reveal that many of the farmers' practices had rationality.

2.6. INTEGRATION OF SCIENTIFIC AND INDIGENOUS WISDOM

Rudramoorthy (1964) pointed out that a judicious combination of folk knowledge and scientific knowledge will help speed up the adoption of improved practices by the farmers.

Warren (1987) pointed out that by comparing and contrasting Indigenous knowledge Systems with the scientific technologies of International Agricultural Research and Development Centers (IARDCs) and Regional Research Stations, it is possible to see where technologies can be utilized to improve upon local systems.

Chambers (1988) revealed that the integration of needs and wants analysis with development work represents a challenge for selection and application of appropriate technology, as does the use of traditional or indigenous knowledge. Indigenous knowledge is seen as having the effect of complementing science – based knowledge in finding solutions to development problems

Sustainable agriculture lies in the modern technology which should be followed along with traditional technology in a harmonious fashion so that agriculture could remain sustainable which is today's demand (De walt, 1994; Altieri, 1996; Lal and Singh, 1997; Obinne and Ozowa, 1997; Gupta, 2000; Chandra *et al.*, 2000).

Development thinkers argued that IK of farmers may not work in every situation, but the integration of local and external technologies can result in appropriate solutions (Shah, 1994).

Suzuki *et al.* (1992) reported that differences between indigenous knowledge and western "Scientific" knowledge, and stress that the future of the human species on the planet depends on a joining of these two fundamentally different world views. While the gap between these two is large, there is common ground and through a series of case studies, each of which represents a particular ecological theme, the authors compare scientific and aboriginal explanations of natural phenomena. This comparison is meant to suggest both the mortality and validity of indigenous views of nature and to juxtapose the knowledge and world views of native and western "Elders".

Agrawal (1995) revealed that both the concept of indigenous knowledge and its role in development are problematic issues as currently conceptualized. To productively engage indigenous knowledge in development, we must go beyond the dichotomy of indigenous Vs Scientific and work towards greater autonomy for 'indigenous' peoples.

Abstraction and conceptualization of IK and its integration with modern farming techniques can evolve efficient resource management system (Sain *et al.,* 1995; Sulaja, 1999).

Pastakia (1996) reported that modern science does not merely help explain the functional viability of the technology, but also provides a basis for abstraction and generalization. Modern science can blend in with the traditional methods while leaving other choices intact. Farmers can be extremely creative in solving local problems. But the issue is whether their knowledge systems can be blended with formal scientific research. One block may possibly be the tension between the farmer's interest in solving the problem and the scientists' interest in developing a new theory.

Several cases have been reported where traditional practices have surpassed modern technical know how (Verma *et al.,* 1997; Tripathi *et al.,* 1997). The integration of scientific and indigenous wisdom would help develop need based technologies.

Mishra (1998) reported that it is made possible for the integration of the indigenous knowledge of local communities with the 'Scientific' principles involved in modern forestry management and led not only to better management of forests, but to a richer anthropogenic activity and better biomass yields.

Ramkumar (1998) opined that if the livestock sector in India is to be improved and sustained, the value of ITK and farmer formulated rations must be recognized and efficiently blended with scientific technology or information.

Abigail Andah (2000) revealed that an improved indigenous technology is the result of the combination of modern technological practices and traditional food processing practices. The traditional character of the final product may be maintained, modified and changed depending on the target market for the final product. Upgrading of indigenous technology helps to diversify the type of products that could be derived from food commodities through the development of new or novel food products. New product development also helps to substitute existing ingredients with more cost – effective and readily available ones. This may lead to import substitution and foreign – exchange savings. He further pointed out that best practices promote the use of energy – efficient technologies that use less resource and create less environmental degradation. The up graded technology should result in increased productivity to reduce labour and costs through faster production and lower production costs.

According to Kimmerer (2000), native peoples' traditional knowledge of the land differs from scientific knowledge; both have strengths that suggest the value of partnership between them.

A critical analysis of existing traditional agricultural wisdom, its element and concept and need to integrate the same with scientific wisdom for boosting productivity and its sustainablitli8ty seems to be the demand of the era (Singh *et al.,* 2001).

Lakshmanan (2000) and Pinstrup *et al.* (2000) stated that modern technological means and traditional experience of the past when blended together can definitely help evolving nationalist programme which can help increasing the annual food production of India.

To summarize, there is immense scope for blending the traditional practices with modern techniques to make farming sustainable.

2.7. CONCEPT AND ADVANTAGES OF PARTICIPATORY RESEARCH AND PARTICIPATORY TECHNOLOGY DEVELOPMENT (PTD)

In Participatory research, researchers and farmers are the two equal partners of a collaborative interface (Biggs, 1989).

Wallace – Bruce (1991) reported that Participatory Technology Development approach where technologists listen to, work together with and build on the knowledge of women in developing an efficient technology.

FAO (1992) reported that one of the important conclusions stressed the need for participatory approach to ensure that vital local knowledge is recognized and tapped in development. Local people should be recognized as managers knowledgeable about local resources, the environment and their operations. Consensus building through group participatory approach should be adopted as the strategy for assisting with development.

Haverkort *et al.* (1992) stated that in the process of technology development, knowledge of indigenous livelihoods is an indispensable resource and further pointed out that Indigenous Knowledge reflects the dignity of the local community and puts its members on an equal footing with the outsiders involved in the process of technology development.

Talwar and Singh (1992) have reported the use of participant techniques in data collection of indigenous knowledge on climate and crop pest climate interaction.

National Council of Applied Economic Research (NCAER) (1993) concluded that the cost incurred for the training of field staff and data collection for the sample survey- based study, were higher by more than one half of the cost incurred in PRA based data collection.

Farmers' innovations often serve as a basis for new research or for modifying recommendations (Stroud 1993; van Veldhuizen *et al.* 1997).

The superiority of farmers participatory research over the conventional research based approach was that, the technology developed through farmer participation was better adapted to local conditions (Worade and Mekhib 1993; Prema *et al.,* 2000).

The approach would be extremely flexible, meaningful and joyous to the participants and researchers, provided the later believed that the people were the store houses of knowledge and had clear perception of their own needs (Ahamed *et al.,* 1996).

Quist (1996) supported the intension that farmers were not exposed to increased risk when involved in participatory research. This is because the farmers tend to 'test' the new material variety on small often inferior plots before fully adopting that particular variety.

When IKS and PRA approaches are placed on the context of communicative action theory, they will gain theoretical, political and practical significance and should improve the cross-cultural co-operation for development (Hess, 1997).

The potential of IKS as a foundation for sustainable development has established participatory methodology as a tool to compile and explain the scientific reasoning and adaptability of ITKs (Kashyap *et al.,* 2000).

Singh *et al.* (2001) analyzed critically the full involvement of the farmers / farm women by way of using different tools of participatory Rural Appraisal (PRA) for documenting the ITK items to find out the rationale behind it.

2.7.1. Limitations of Conventional Surveys

Formal surveys are based on the problems associated with sampling, reaction of the respondents, and survey staff, inaccuracy in recording and interview errors (Moris, 1970)

Bardhan (1989) criticized that in conventional surveys there could be persistent loss of information in case of ambiguities in the interpretation of terms and categories with overlapping boundaries.

Chambers (1991) observed that data collection in large questionnaire surveys could be costly, time consuming and most of the data collected lay idle without being used by anybody.

Formal surveys require five to seven times more working days compared to informal surveys and the time needed to publish the results were three times longer for formal surveys (Hubert, 1991).

Gill (1993) reported that the problems of official statistics based on surveys that many of the statistics were characterized by unreality, gaps, over – aggregation, inaccuracies and mutual inconsistencies. Untimely reporting and policy formulation based on such incorrect empirical data lead to wrong decision making.

Comparing the cost and time of PRA methods and conventional surveys, the National Council of Applied Economic Research (NCAER – 1993) concluded that the cost incurred for the training of field staff and data collection for the sample survey based study, were higher by more than one – half of the cost incurred in PRA based data collection.

2.7.2. Advantages of Participatory Approach in Data Collection

Brokensha *et al.* (1980) stated that Participatory Development or "Development from below" must be built on effective communication (Working partnership) with local groups and on constructive use of local institutions: and working partnerships entail, in part, exploring the complementary relationships between indigenous and "Scientific" knowledge.

Ashby *et al.* (1987) described ways in which farmers evaluated the potential of different varieties developed by the Scientists. It is not just the judgements that one can learn from Participatory exercises; the opportunity to learn about the criteria for making judgements is much more important. She further emphasized that Participatory research should involve farmers as co – investigators and researchers and demonstrated, through farmer – managed trials, creative ways of understanding farmers criteria for selecting varieties.

Gupta (1990) pointed out that Indigenous Knowledge systems form the basis for informal experimentation of farmers. The factors which influence farmer experimentation are (1) ecological: Innovations that result due to interaction among crops, soil and climate. (2) Historical: a majority happening such as crop failure or year of glut or scarcity. (3) Serendipity: a practice discovered by farmers accidentally.

(4) economical: farmers innovate new practices taking advantage of government subsidies for flood and drought relief activities.

Haverkort and Zeenw (1991) and *Okali et al.* (1994) highlighted Farmer Participatory approach as an empowering process as "those with a legitimate interest in the outcome of research are able to exert some influence on priorities and decisions."

There is a growing awareness that agricultural research and development must build upon farmer expertise; identifying, facilitating and building upon local innovation. As the Transfer of Technology (ToT) paradigm is increasingly replaced in African agriculture by the drive to facilitate Participatory Technology Development (PTD) (Haverkort *et al.,* 1991; Martin and Sherington, 1997), it is becoming evident that scientists and extensionists need to develop a more facilitator role.

Warren (1992) stated that Indigenous Knowledge systems also provide mechanisms for facilitating understanding and communications between outsiders (extensionists, researchers) and insiders (farmers). Improved understanding and communications enhance participatory approaches to problem identification.

Studies by Worade and Mekhib (1993) and Prema *et al.* (2000) proved the superiority of farmer participatory research over the conventional researcher based approach, as the technology developed thro-ugh the farmer participation was better adapted to local conditions.

Andah *et al.* (1994) reported that there has to be constant forward and backward flow of information between key players in the process of technology development so as to achieve a balance between 'Science – push and demand pull'. There is need for close interaction between all persons concerned with decision making in technology development, extension and diffusion. A participatory approach that requires the active involvement of all stakeholders, with the target group at the centre, facilitates the selection of appropriate choices and contributes to eventual success of the project. They further stated that the use of locally available raw materials, coupled with local skills for the design and fabrication of equipment and machinery have proved to be essential for sustainability.

Bentley *et al.* (1994) have argued, there are many areas where scientists can help farmers by providing information and technologies to which they have not had access earlier. This strengthens the case for true participatory research between scientist and farmer.

The Indian Council of Agricultural Research (ICAR 1995) stressed the need for developing appropriate research programmes through farmer participation in order to account for complexities of required knowledge involved and the ecology as well as social environment in which the farmers work.

Ahemad *et al.* (1996) opined that the participatory approaches were extremely flexible, meaningful and joyous to the participants and researchers, provided the later believed that the people are store-houses of knowledge and have clear perception of their own needs.

There was a good agreement between farmers' perception of the variety and crop harvest results, when the same varieties were grown on the field and tested (Sthapit *et al.*, 1996).

Warren *et al.* (1996) indicated that local planning and management systems, local levels of technology and development and community – based systems of evaluation and capacity – building.

Okali *et al.* (1997) stated that farmers are involved in ongoing processes of local knowledge creation through site specific learning, which, in the short term, results primarily in small adaptations to farming practice and, in the long term, contributes to the development of new farming systems.

Van Veldhuizen *et al.* (1997) stated that farmers' innovations often serve as a basis for new research or for modifying recommendations.

Abigail Andah (2000) pointed out that best practices promote the use of energy – efficient technologies that use less resource and create less environmental degradation. The upgraded technology should result in increased productivity to reduce labour and costs through faster production and lower production cost.

Estelitta *et al.* (2000) reported that the basic approach is to mobilize and utilize natural resources for the development of locally adaptable technology in a massive way for the sustainable improvement of rice productivity.

Analyzing the various research findings Pal *et al.* (2000) opined that private sector's reluctance to promote varieties for fear of rejection by the farmers could be effectively managed by ensuring farmer participation in varietal selection and release.

Prema *et al.* (2000) opined non adoption of improved rice varieties in Kerala was because of high complex, diverse and risk prone and location specific agricultural production system and lack of participatory endeavors in research and extension.

Weltzien *et al.* (2000); Thiele *et al.* (1997) and Careasco *et al.* (1997) after analyzing the various reports categorically established the role played by social science researchers in identifying and prioritizing farmers criteria in varietal selection to improve targeting of breeding activities.

Kibwana *et al.* (2001) concluded that a modified PTD approach is emerging which does not deal with problem analysis alone but rather seeks to link up with local problem – solving initiatives, thus using indigenous knowledge at the start of innovation development instead of at the end.

Farmer experimentation must be supported, innovators and their innovations identified, and where necessary it may be possible to work with innovators to optimize their innovations, and disseminate them to other smallholders who may benefit from them (Reij and Waters-Bayer, 2001).

Rocheleau (2003) indicated that "thousands of field workers conduct isolated, undocumented research in extension and development programs on forestry, agriculture and conservation." PTD is practiced as an approach to extension, with development workers supporting farmers in learning-by-doing without involving researchers.

Marschke (2003) stated that Participatory Technology Development (PTD) and Farmers Field School became fully accepted approaches for information exchange between farmers and researchers. The use of Indigenous knowledge in innovation development led to a number of surprising successes. Most authors emphasize the importance to hybridize indigenous and formal scientific knowledge.

Sumberg *et al.* (2003) stated that various categories of technological innovations each being more or less apt for participatory inclusion of indigenous knowledge according to the solution space they offer. They distinguish: 'mass market' technologies (e.g., new hybrid variety) small solution space, no adaptive research necessary, system technologies (e.g. agro – forestry system), medium solution space, adaptive research possible, defensive technologies (self – consumption – risk reduction) large solution space, adaptive and participatory research necessary.

Kishore Kumar *et al.* (2004) stated that 66.67 per cent of the farmers had good perception about the PTD concept. Since the farmers were literate and innovative they got sufficient exposure to the new farming practices including PTD. Since they were conceived about the results of PTD in their own fields, they had a good perception about

PTD. And further pointed out that 80 per cent of the farmers participation in PTD was high and that farmers had good level of knowledge regarding PTD concept and its related practices. He further pointed out that PTD is the practical process of bringing together the knowledge and research capacities of the local farming communities with that of the commercial and scientific institutions in an interactive way.

Morris *et al.* (2004) reported that the term "Participatory Plant Breeding" has been used to refer to different forms of interaction between farmers and researchers at different stages of the crop research and improvement process. Participatory crop research and improvement can involve Scientists, farmers, extension agents, industry and consumers and others in the research process. The objective is to facilitate quicker and more extensive uptake of new cropping technologies by shifting the locus of plant Genetic Improvement research toward the local level through direct stakeholder involvement at different stages of the breeding process.

Lilja *et al.* (2005) stated that participatory methodologies can play an important role in identifying and valuing these traits and if this information is fed back into the design and development of new varieties, it can help to make them more relevant and appropriate, so that they generate more benefits to these people

Reed (2007) reported that in order to facilitate Participatory Technology Development (PTD) in African agriculture, extensionists and scientists must collaborate with local innovators to optimise (where necessary) and disseminate their innovations.

2.8. TRADITIONAL TECHNOLOGIES USED FOR AGRICULTURAL PURPOSES

Majumdar (1956) reported that increased growth and yield of banana when ammonium sulphate was applied along with cattle manure as compared with either ground nut cake or mustard oil cake as a replacement for N with the cattle manure

Champion *et al.* (1958) in 'Robusta' banana demonstrated that N supplied in the form of urea – formaldehyde applied once per year gave better yields.

Venkatarayappa *et al.* (1978) reported that the post – shooting application of Potassium di-hydrogen phosphate (2% spray on foliage or bunch) on Cavendish banana increased the volume of the fruits.

Chaman Lal *et al.* (2006) reported that sprinkling of *chula* (a traditionally made fire place for cooking food) ash over and around the vegetable plants and in fields is effective against insect pests viz., beetles, leaf defoliating insects, leaf miners, thrips and aphids. It is the cheapest practice for small farmers. It is simple and easy to apply and effective method for insect pests, use of local raw material, eco – friendly and has no side effect, it enhances crop productivity, compatible with internal resources of household, technically feasible, and widely used.

Chaman Lal *et al.* (2006) further indicated that Cow urine + *Vitex negundo* (Nirgundi) + *Ferula asafetida* (Hing) mixture is considered very effective, eco – friendly insecticidal treatment against insect pests of Paddy and Wheat crop. It is technically feasible, compatible with existing farming systems and internal resources of the household, simple and easy to prepare, controls insect – pest, it enhances crop productivity.

Chaman Lal *et al.* (2006) stated that ash + soil mixture and cow urine are used to protect the cabbage plant from insect pests. It is simple and easy to follow, use of local raw material, helpful to overcome pest problems, it enhances crop productivity, eco – friendly, and checks insects – pests control, compatible with internal resources of the household.

Joshi *et al.* (2006) reported that in hill areas, wood obtained from forest has been major fuel. Ash dust is a product obtained after the burning of fuel wood. The kitchen ash, thus obtained has been mixed with the farm yard manure and applied into the field. This practice is quite common in vegetable crops, especially potato. The organic manure improves the soil structure and water holding capacity in addition to supply of nutrients. The ash dust contains phosphorus to the crop. Moreover, the farmers reported that mixing of ash is helpful in spreading of manure into the field.

Joshi *et al.* (2006) further stated that large size potato tubers have been cut into two or three pieces, depending on the number of eyes (growing buds) on the tuber. This practice saves the potato seed. The cut potatoes when placed in the soil have been exposed to the pest present in the soil. To prevent it, farmers apply ash on the cut surface of the potato. The cut potatoes have been taken in a *Daliya* and ash has been spread over it and mixed properly. The ash makes a larger over the cut surface. This layer act as a physical barrier for the pest present in soil. As the ash has been obtained from burning of wood it has been free from any type of micro organisms.

Joshi *et al.* (2006) indicated that ash obtained from burning of fuel wood has been mixed with the food grains. The probable reason for its use may be its ability to absorb moisture, which otherwise will be absorbed by the grains and make them prone to pest attack. Moreover, ash acts as a physical barrier for movements of pests from one grain to another.

Karthikeyan *et al.* (2006) indicated that farmers of dry land agriculture faced the problem of stickiness of cotton seeds creating difficulty during dibbling of the seeds. To overcome this problem, cotton growers follow a practice of seed treatment with cow dung slurry locally called as *Sanipal pidithal*. Nearly 60% farmers adopt this technology as extra cost is involved. It is both safer and cheaper method to facilitate sowing operation. It is moderately effective (60%) in sowing seeds. They further added that chilli growers use cow dung for the treatment of chilli seeds. Nearly 40% farmers adopt this technology as no extra cost is involved. This process is effective (80%) in inducing the germination of seeds and control of seed borne diseases like fruit rot. To avoid water stress during the crop growth, an indigenous seed hardening method using cow's urine is practiced. Cow's urine when treated with Ragi seeds were found to be effective against seed borne diseases like smut and also in inducing drought resistance in crop plants. Nearly 60% farmers have adopted this low cost technology.

Senthil Vinaygam *et al.* (2006) reported that mixing of seed with ash is adopted by 22% of respondents. About 39% of researchers indicated that treating seed with ash, which is an inert material provides protective physical covering against pest and disease inoculums. The ash is known to be an excellent source of potash, which is a quality element and protects crops against water stress as well as pest and diseases. Moreover, the moisture availability around seed is enhanced leading to faster germination and growth. The study also revealed that dusting of cow dung ash on the infected leaves acts as repellant and antecedent to control beetles as reported by 20% of the respondents. Two kg of cow's ghee, 2 kg of honey, cow urine and 1 bag of FYM is used for control of insect and diseases. This practice was reported by 19% of the respondents. They further stated that more than 50% of the respondents followed the practice to control the pest. The rationale has already been explained in the preceding text. Again, the only limitation for its effectiveness is repeated/frequent sprays and adoption on community basis. Such methods help to increase the population of natural enemies of insects. Spray of distillate from neem, *karanj* and *Ipomoea* leaves dipped in 20 litres of cow urine aged in earthen vessel kept underground for 3-4 weeks in 200 litres of water is used for protecting the soybean crop from caterpillars. From the acetone soluble fraction of the seed of *karanj,* a furo-coumarin compound is isolated to protect grains from the pests as indicated by 38% of the researchers.

Baldeo Singh *et al.* (2007) reported that to provide continuity to grassroots experimentation and evolution of indigenous technologies it is imperative to transfer science among farmers and also develop their experimentation capabilities. Endeavour in this direction was made through farmer participatory validation of indigenous technologies (management of khaira disease in rice with lime and cow dung, and management of termites in sugarcane with lime and salt. Farmers of Bareilly and Badaun districts of Uttar Pradesh used mixture of two or three kilograms of lime and one bucket of cow dung mixed in 200 litres of water on the affected crop of paddy for management of khaira disease. On – farm validation trials were laid out using RBD experimental design with five treatments (control – T_1; spray of lime (3 kg) + cow dung (1 bucket) mixed in 200 litres of water – T_2; spray of lime (0.25%) + zinc sulphate (0.5%) – T_3; spray of zinc sulphate (0.5%) alone – T_4; soil application of lime (35 kg/ha) + sodium chloride (75 kg/ha) – T_5. After the application of treatment, incidence of khaira disease reduced in all the cases but at varying degrees. Symptoms nearly disappeared in case of treatments T_3 and T_4. However, in case of treatment with ITK, reduction in the symptoms was relatively low.

He further reported that farmers in Bareilly and Shahjahanpur districts of Uttar Pradesh followed soil application of 2.5 kg lime and 5 kg salt to control termites in sugarcane. Four on – farm validation trials one at Dabhora and three at Manda village of Bareilly district were laid out with pre – sowing treatments of lime (39 kg/ha) and salt (78 kg/ha), Lime (39 kg/ha) and salt (78 kg/ha), salt alone (78 kg/ha), and phorate (10 kg/ha) as chemical control measure. Apparently, the findings showed that termite infestation was managed by application of salt along with lime as well as alone to some extent though not at par with phorate.

Kotur *et al.* (2007) reported that enriched cow dung applied to the de-navelled distal end of bunch of Robusta banana earns additional income when the excised male bud is used as vegetable.

He further stated that reported that in demonstration taken up in the field, a response of 22% increase in bunch weight was observed in the Robusta banana plantation of Shri. Umesh of Uddichikkanahalli village. In Ney Poovan banana plantation of Shri Hombe Gowda of Bidadi Village near Bangalore a response of 28% bunch yield was evident. For those farmers who wish to practice organic farming, blending urine of cow with the dung and applying to the distal end will be beneficial as 500 gm of cow dung contains substantial amounts of about 5.5 gm of N, 3.5 gm of K, 1.6 gm of S besides other mineral and bio chemicals beneficial to the enhancement of size of the banana fingers. The technique involves blending 15 g (approximately 7.5 gm of urea, and 7.5 gm of sulphate of potash) dissolved in 100 ml water in 500 gm of fresh cow dung and applying the slurry to the de – navelled stalk – end of Robusta bunch soon after fruit set. Experimentally it was found that by this treatment, the bunch weight increased by 67% over control in which the male flower was retained till harvest.

He further reported that in Ney Poovan banana, blending 5 gm of ammonium sulphate 9 or 2.5 gm of urea) and 2.5 gm of sulphate of potash and applied similarly showed a response of 78% by absorbing 42% of N applied. The quality of banana was not affected in any way by these treatments. The technology is much cost – effective since the cost involved is about Rs. 2/- per bunch while the improvement is of 4-5 kg of Robusta and 2-3 kg of Ney Poovan banana bunch.

RESEARCH METHODOLOGY

Appropriate research methods and techniques were adopted according to the objectives set forth for the study. The research procedure followed is presented in this chapter under the following sub headings:

3.1. Design of the Research

3.2. Phase I of the Study

3.3. Phase II of the Study

3.4. Procedure employed in data collection

3.5. Operationalization and measurement of attribute measures of the treatments in banana

3.6. Statistical tools used in the study

3.7. Operationalization of concepts and definitions

3.1. DESIGN OF RESEARCH

3.1.1. Design of Participatory Technology Development (PTD) through Field Experimentation

Based on the analysis of available literature and keeping in view the objectives of the study, it could be well inferred that the first phase of the study is field experimentation. According to Kerlinger (1964) Field experimentation is a research study in a realistic situation in which one or more independent variables are manipulated by the experimenter under as carefully controlled conditions as the situation will permit.

The second phase of the study inferred that most of the attributes included were ex-post facto in nature and offered little chance to be manipulated by the researcher. Therefore, ex-post facto research design was considered appropriate to bc used for the study.

According to Kerlinger (1964) ex-post facto research is a systematic empirical enquiry in which the researcher does not have direct control over the independent variables because their manipulation have already occurred or because they are inherently not manipulable.

Design of Participatory Technology Development (PTD) through Field Experimentation was based on the modality of Participatory Research. Participatory Research being heuristic in nature it provided more opportunity to change the mode of operation to further improve the programme by overcoming the pit falls, if any, identified during the previous experiences.

Participatory Research is a continuous cycle in which insiders and outsiders together decide what needs to be researched, design the research, and collect the necessary information. Insiders and outsiders are involved in setting research priorities and designing research plots. (Winter, 1989)

Participatory methodologies aim to institutionalize a role for farmers in deciding what technology to test, how to test it, what to feed back and what recommendations to make. Instead of blanket recommendations, farmers take part in selecting promising items and participate in taking observations and data for evaluating the results (Prema *et al.,* 2000)

The approach in the present study was to increase the involvement of farmers not only in the physical implementation of the experiments but also in the definition of research needs, the design and evaluation of research results in these specialist knowledge. The study was conducted in two phases. The first phase accomplished the major objective of conducting field experiment in participating farmer's fields in K. Dharmathupatti and Kondayampatti areas. The second phase focused on the rest of the objectives viz., Degree of dependability in treatments in Banana, Adoption of treatments in Banana, Assignment of weightage score to treatments in Banana, Evaluative perception of treatments in Banana and Rationalization of treatments in Banana by Key Informant Farmers (KIFs), extensionists and scientists with respect to the treatments for further validation.

3.2. PHASE I OF THE STUDY

The field experiments were conducted during January to May 2008 at Kondayampatti of Vadipatti taluk of Madurai and K. Dharmathupatti of Oddanchathiram taluk of Dindigul District of Tamil Nadu to study the influence of post - shooting application of various ingredients to enhance uniform growth of bunch from top to bottom in Banana. The profile of the research areas are as follows:

3.2.1. Profile of the Research Area

3.2.1.1. Madurai District Profile

Madurai District is situated in the Southern part of Tamil Nadu state bordering Dindigul and Thiruchirapalli districts in the north, Sivagangai on the East, Theni on the west and Virudhunagar district on the south. The district's geographical position is in such a way that it is surrounded by several mountains. The Madurai city has 3 hills as its city boundary. Yanaimalai, Nagamalai, Pasumalai named after elephant, snake and cow respectively. It is famous for Jasmine flowers. Jasmine flowers are transported to other cities of India from Madurai.

3.2.1.1.1. Facts and Figures

Area	3741.73 sq. km.
Latitude	9°30'and 10°30' N
Longitude	77°and 78°30' E
Population (2001)	25,62,279
Males	12,95,124
Females	12,76,155
Population density	733 (Per sq. km.)
Literacy Rate	70%
No. of Sub-Division	02
No. of Tehsil	07
No. of Blocks	13
No. of Villages	669
Average rainfall	214.25 mm

3.2.1.1.2. Agricultural Details

Basic details

1. Farming Population (as per 2001 Census)	
Small Farmers	58,716
Marginal Farmers	2,05,771
Agricultural Labourers	4,01,867

Total farming population	**6,66,354**
2. No. of Agricultural Divisions	6 Nos.
3. No. of Agrl. Extension Centres to cater farmers needs of Agricultural inputs and technical advices.	
a) Main Agrl. Extension Centres	13
b) Sub Agrl. Extension Centres	18
4. Seed Processing Unit	2 Nos.
5 State Seed Farm	1 No. at Vinayagapuram, Melur
6 Soil Testing Laboratory	2 Nos. at Madurai
7 Fertilizer Testing Lab	1 No. at Madurai
8 Pesticide Testing Lab.	1 No. at Madurai
9 Sugarcane Parasite Breeding Centre	1 No. at Vinayagapuram, Melur
10 Coconut Parasite Breeding Centre	1 No. at Madurai
11 I.P.M. Lab	1 No. at Vinayagapuram, Melur
12 Bio Lab	1 No. at Vinayagapuram, Melur
13 Agriculture College and Research Institute (Tamil Nadu Agricultural University Control)	1 No. at Othakadai, Madurai.

Irrigation facilities

1. Dams - irrigating areas, Madurai District	3	Periyar Vaigai Sathiyar
2. Irrigation channels / canals		51 Nos. (169.6 K Ms.)
3 Tanks		2579 Nos.
4 Wells		44,698 Nos
5. Periyar - Vaigai Ayacut areas		
(a) Double crop		13860 Ha.
(b) Single crop		49500 Ha.

Major crops and their normal area

Sl. No.	Crops	Normal Area	Productivity M.T./Ha.	Production L.M.T.
1	Paddy	80000	4.400 (Rice)	3.520 (Rice)
2	Millets	16700	1.200	0.196
3	Pulses	24,000	0.700	0.155
4	Cotton	13,100 .	2.5 Bales	29,900 Bales

5	Oilseeds	15,100	0.900	0.135
6	Sugarcane	7,000	200	1.400

Soil Type, pH and Ground water colour

S.N.	Block	Soil Type	pH	Ground water colour
1	Madurai East	Clay loam	7 - 8	White
2	Madurai West	Clay loam	7 - 8	White
3	Tirupparankundram	Red loam, Black cotton	6.5 - 8.7	White
4	Melur	Red loam	6.7 - 7.5	White
5	Kottampatti	Red loam	6.7 - 7.5	Grey
6	Vadipatty	Red loam, clay loam	6.7 - 7.5	Black
7	Alanganallur	-do-	6.7 - 7.5	Black
8	Tirumangalam	Red loam & Black cotton	6.7 - 7.9	White
9	Kalligudi	-do-	6.7 - 7.9	White
10	T. Kallupatti	-do-	6.7 - 7.9	Black
11	Usilampatti	-do-	6.7 - 7.9	Black
12	Chellampatti	-do-	7.5 - 8.0	Black
13	Sedapatty	-do-	7.5 - 8.5	Black

3.2.1.2. Dindigul District Profile

Dindigul, which was under the rule of the famous Muslim Monarch, Tippusultan, has a glorious past. The historical Rock Fort of this district was constructed by the famous Naik King Muthukrishnappa Naicker. It is located between 10°05" and 10°9" North Latitude and 77°30" and 78°20" East Longitude. This district is bounded by Erode, Coimbatore, Karur and Trichy districts on the North, by Sivaganga and Tiruchi District on the East, by Madurai district on the South and by Theni and Coimbatore Districts and Kerala State on the West. It is spread over an area of 6266.64 Sq. Km. It comprises of 3 Revenue Divisions, 8 Taluks and 14 Panchayat Unions, According to 2001 Census, its population is 19,23,014.

For a long time, Dindigul town has been associated with Iron locks, Iron safe of good quality and durability. A lock manufacturing unit under co-operative sector is functioning here. Another industry for which Dindigul is noted is Leather Tanning. This district is having a flourishing handloom industry at Chinnalapatti, which is located at

11 kms away from Dindigul on the Madurai-Dindigul road. Art-Silk saris and Sungudi saris produced in Chinnalapatti are famous through out India. More than 1000 families are engaged in this Industry.

Dindigul city which is an important wholesale market for Onion and Groundnut has the network of inter-district roads connecting Coimbatore, Erode, Tiruchi, Karur, Madurai and Sivaganga District. Educationally, Dindigul is a well developed and popular city. It has got many High and Higher Secondary Schools to its credit.

3.2.1.2.1. District profile at a glance

GEOGRAPHICAL POSITION	
North Latitude	Between 10^0 05' and 10^0 09'
East Longitude	Between 77^0 30' and 78^0 20'
AREA AND POPULATION	2001 Census
Area (Sq.Km)	6266.64
Population	19,23,014
Density	306
TEMPERATURE (in Celsius)	
Plains	34.3(Max), 22.5 (Min)
Hill Stations	22.0 (Max), 8.0 (Min)
RAINFALL (in mm)	
North East Monsoon	741.2
South West Monsoon	247.5

3.2.1.2.2. Agricultural Details

Total Cultivated Area (Ha.)	2,61,758
Net Area Sown (Ha.)	253505
Area Sown more than once (Ha.)	8253

	Area (Ha.)	Production (Tonnes)
Paddy	23735	8378
Millets and Other Cereals	81610	11011
Pulses	27135	37310
Sugarcane (Gur)	7014	8136
Groundnut	22070	580441

Gingelly	1459	655091
Cotton (BL)	1999	5657

Agricultural Land Holdings	Holdings	Area	Average Size of Holdings
	302711	334835	1.11.0

IRRIGATION	
I. Net Area Irrigated by (Hec.)	
a. Government Canals	7901
b. Private Canals	--
c. Tanks	3603
d. Tube Wells	2368
e. Open Wells	89847
f. Other Sources	973
Total Net Area Irrigated	104672
II. Gross Area Irrigated	**112071**

RAINFALL

The annual average rainfall (50 years) in 836.0 mm excluding Kodaikanal taluk. The season wise rainfall is given below:

(1) Winter period (January – February)	-	44.7 mm
(2) Summer period (March – May)	-	155.1 mm
(3) South West Monsoon (June – Sept.)	-	218.3 mm
(4) North East Monsoon (Oct. – Dec.)	-	417.9 mm
	Total	**836.0 mm**

CROPPING PATTERN

(a)Wet lands

I	Crop	:	Paddy (Oct. – Feb.)
II	Crop	:	Paddy (Feb. – March)
I	Crop	:	Paddy (Oct. – Feb.)
II	Crop	:	Pulses (March. – May)
	Single Crop	:	Sugarcane / Banana (Dec.–Jan.) (June–July)

(b) Garden Lands

I Crop	:	Paddy (Aug – Dec.)
II Crop	:	Millets / Pulses / Groundnut/ Vegetables (Dec, Jan – March)
I Crop	:	Cotton (Sept. – Feb.)
II Crop	:	Millets / Pulses / Vegetables (Feb. – March)
Single Crop	:	Sugarcane / Banana (Dec. – Jan.) (June – July)

(c) Dry lands

(i) Red soils : Cholam (June – Aug.) Cumbu (Aug.–Sept.) Groundnut (June – July); Minor Millets; Horsegram (Sept. – Oct.)

(ii) Black soils : Cholam (Sept.–Oct.); Cotton (Sept.–Oct); Sunflower (Sept. – Oct.)

Normal Area, Production and Productivity of Important Crops

Sl. No	Crop	Area (Ha)	Production (MT)	Productivity (kgs/ha.)	Average Yield (Kgs/Ha)
1	Paddy	25000	100000*(Rice)*	4000*(Rice)*	4680
2	Millets	112000	112000	1000	1465
3	Pulses	45300	27180	600	563
4	Cotton (Bales)	8000	19200	2.4	2.93
5	Sugarcane (Gur)	5500	66000	12 MT *(Gur)*	12.56
6	Oilseeds	36200	50680	1400	1680

LAND UTILISATION

Classification	Area (Hectare)
Forest	138923
Barren and Uncultivated lands	36210
Land to Non-Agricultural Uses	65013
Cultivable Waste	8846
Permanent pastures other Grazing Land	6946
Land under miscellaneous Tree crops and Gross not included in Net Area sown	6491

Current Fallows	19752
Other Fallow Land	123933
Net Area sown	220550
TOTAL GEOGRAPHICAL AREA	626664

3.2.2. Selection of Locations

K. Dharmathupatti village of Reddiarchathiram Block, Dindigul District and Kondayampatti village of Alanganallur block, Madurai District is purposively selected for conducting the field experiments based on more area under banana cultivation. The banana belt areas were selected to conduct key informants workshop. In key informants workshop the most co-operative and responsive areas with progressive banana growers like K. Dharmathupatti and Kondayampatti villages were selected for field experiments.

3.2.3. Selection of Technology

The Project entitled "Documentation and Preservation of Agricultural Traditional Knowledge through Farmer Participatory Approach" undertaken by the Department of Agricultural Extension & Rural Sociology, Agricultural College & Research Institute, Madurai had identified and documented the indigenous technology that the farmers in Kannivadi area of Tamil Nadu used viz., house hold wood ash and cow urine in Banana at the time of last flower cutting in order to get the uniform growth of bunches from top to bottom. The same was taken for the present research.

3.2.4. Selection of Crop and Variety

The crop Banana, a shrub is selected for the study. Banana is the common name for a fruit and also the herbaceous plants of the genus *Musa* which produce the commonly eaten fruit. Banana plants are of the family Musaceae. They are cultivated primarily for their fruit, and to a lesser extent for the production of fibre and as ornamental plants. As the bananas are mainly tall, upright, and fairly sturdy, they are often mistaken as trees, when the truth is the main or upright stem is called a *pseudostem*, literally meaning "fake stem", which for some species can obtain a height of up to 2–8 m, with leaves of up to 3.5 m in length. Each pseudostem can produce a bunch of yellow, green, or even red bananas before dying and being replaced by another pseudostem.

The banana fruits grow in hanging clusters, with up to 20 fruit to a tier (called a *hand*), and 3-20 tiers to a bunch. The total of the hanging clusters is known as a bunch, or commercially as a "banana stem", and can weigh from 30–50 kg. The fruit averages 125 g, of which approximately 75% is water and 25% dry matter content. Each individual fruit (known as a banana or 'finger') has a protective outer layer (a peel or skin) with a fleshy edible inner portion. Both skin and inner part can be eaten raw or cooked. Western cultures generally eat the inside raw and throw away the skin while some Asian cultures generally eat both the skin and inside cooked. Typically, the fruit has numerous strings (called 'phloem bundles') which run between the skin and inner part. Bananas are a valuable source of vitamin B_6, vitamin C, and potassium.

G-9, Robusta, Poovan and Red banana varieties of Banana crop were chosen for the experiments in the research as detailed below.

Area	**Field – 1 - Variety**	**Field – 2 - Variety**	**Field – 3 - Variety**
Kondayampatti of Madurai District	G - 9	G - 9	G – 9
K. Dharmathupatti of Dindigul District	Robusta	Poovan	Red banana

3.2.5. Execution of the Experiments

Prior to the execution of the experiments, the participating farmers of each location were oriented to the research programme with a list of agenda indicating classification of roles and expectations of farmers and researcher, description of the objectives and purpose of the research. Weekly group meetings were scheduled with advice and active collaboration of local organizations and Tamil Nadu Agricultural University to reemphasize the necessity of the involvement of the selected farmers in the research. Good rapport was built and maintained with farmers by considering all the socio – psychological aspects concerned. At every stage care was taken to make the farmers feel that it was their programme and their full participation and involvement was indispensable for the successful completion of the research.

A group was formed in each location and voluntarily three farmers were selected to conduct field experiments in their field according to the objectives.

3.2.6. Experimental Design and Layout

The experiment was laid out in a Randomized Block Design and replicated thrice. Randomized Block Design is the most widely used in on – farm experimentation for almost all research objectives. Its primary feature was the employment of blocks of more or less equal size, each of which was distributed randomly. This was required for the collection of agronomic data as well as for socio – economic information. It helped to achieve a good representation of different farmers' views and of the usually heterogeneous environmental and management conditions.

3.2.7. Identification of the Combinations through Key Informants Workshop

Key Informants workshops were conducted with the progressive banana farmers and the scientists in the Banana belts of K. Dharmathupatti, Kondayampatti and Melakkal to finalize the treatments with farmer participatory approach. The following 12 treatments were finalized at the end of the key informant's workshop.

3.2.8. Treatments

The treatment details were:

T_1	Household wood ash + Pregnant cow urine
T_2	Brick chamber wood ash + Pregnant cow urine
T_3	Panchakavia + Brick chamber wood ash
T_4	Panchakavia + Household wood ash
T_5	Urea
T_6	Urea + Sulphate of potash
T_7	Fresh cow dung extract + Zinc sulphate + Borax
T_8	Fresh cow dung extract + Zinc sulphate
T_9	Fresh cow dung extract + Borax
T_{10}	Urea + Sulphate of potash + water + Fresh cow dung
T_{11}	Cow dung extract solution
T_{12}	Control

The aforesaid treatments were used for post–shooting application and applied by the following way. The distal axis (male axis) was cut 15 cm away from the last hand fifteen days after the last hand was opened and after removal of the distal axis along with the navel bud (male bud) the aforesaid treatments ingredients were tied in a polythene bag.

3.2.9. Harvest

Harvesting was done uniformly at full maturity stage by eye observation (after the angular growth of the fruit skin disappeared and the colour turned from more or less dark green to light green). The peduncle was cut at 22.5 cm above the first hand as commonly practiced by the growers. The mother plant was cut down by leaving about one meter of the pseudo stem above the ground level.

3.2.10. Biometrical Observations

Biometrical observations on the yield parameters under each treatment replication wise and of bunch were done by following methods of Robin (1985). Data on the yield parameters were recorded at the stage of harvest.

3.2.10.1. Bunch Characters

3.2.10.1.1. Yield t/ha

Yield was calculated based on the total number of fingers per bunch and the spacing of the particular varieties of the each treatment and converted into tonnes per hectare.

3.2.10.1.2. Bunch Weight

Weight of the individual matured bunches was taken in Kg and recorded.

3.2.10.1.3. Number of Hands and Fingers

Total number of hands per bunch and fingers in each hand were counted and recorded.

3.2.10.1.4. Average Weight of the Fruits (Fingers)

Three fruits from the upper row and two from the lower row from the middle part of the second and last hands were selected separately for recording the mean weight (in gram) of the fruits and also for other physical characters (Gottreich *et al.,* 1964). The average finger weight was found out by dividing the bunch weight by total number of fingers in the bunch.

3.2.10.1.5. Volume of the Fruit (Finger)

The volume of the fingers was found out by using the following formula (Simmonds, 1953):

$$V = 0.3537 + 0.0616\ LC^2$$

Where, V = Volume of the finger in cc.

L = Length of the finger in cm.

C = Circumference of girth of the finger at central portion in cm.

3.2.10.1.6. Density (Volume/weight ratio) of Fruit (Fingers)

The density was worked out by dividing volume of finger by weight of each finger and the mean values were recorded for each set of samples.

3.2.10.1.7. Average Length of the Fruits (Fingers)

Finger lengths were measured by a flexible measuring tape from the base of the pedicel to the tip and on a particular side and the average was expressed in centimeters.

3.2.10.1.8. Girth of the Fruits (Fingers)

The girth of the fingers at mid portion was measured and the mean was expressed in centimeters.

3.2.10.1.9. Total Soluble Solids (T.S.S.)

The T.S.S. was determined from fresh pulp by a Zeiss hand juice Brix Refractometer at ambient temperature.

3.2.10.1.10. Total Number of Fingers/Bunch

Total number of fingers per bunch were counted and recorded.

3.3. PHASE II OF THE STUDY

3.3.1. Selection of Respondents

Representatives from all the sub systems identified viz., Farmer Sub System (FSS), Extension Sub System (ESS) and Research Sub System (RSS) were included as respondents for the investigation.

3.3.2 Farmer Sub System (FSS)

A sample of 30 Key Informant Farmers (KIFs) formed the Farmer Sub System (FSS) who participated in the Farmers and Scientists Results sharing meet which was conducted in the Kondayampatti of Madurai District and K. Dharmathupatti of Dindigul District were selected purposively for this study.

3.3.3. Extension Sub System (ESS)

The Assistant Director of Horticulture and Horticultural Officers from Madurai and Dindigul District constituted the respondents of the Extension sub System (ESS). Thus 30 Horticultural Extensionists were selected purposively for this study.

3.3.4 Research Sub System (RSS)

A sample of 30 Agricultural Scientists from Tamil Nadu Agricultural University, and National Research Centre for Banana, Tiruchirapalli were purposively selected on a multidisciplinary basis. (Agronomy, Plant Protection, Horticulture, Spices and Plantation Crops, Plant Physiology, Plant Breeding and Agricultural Extension).

3.4. PROCEDURE EMPLOYED IN DATA COLLECTION

3.4.1. Selection of Attribute Measures of the Treatments under Study

Based on the objectives and by reviewing relevant literature and discussion with experts, the following broad attribute measures were selected: Degree of dependability in treatments in Banana, Status of adoption of treatments in Banana, Assignment of weightage score to treatments in Banana, Evaluative perception of treatments in Banana and Rationalization of Treatments in Banana. Besides Perception towards Participatory Technology Development (PTD) were elicited from all the three systems. The measures of the evaluative perception of treatments in Banana included simplicity/complexity, compatibility with the existing farming system, practicability, availability of raw materials, precision/skill requirement, drudgery, labour intensity and operational costs. In the present study the respondents were asked to express their evaluative perception of each treatment based on the overall consideration of the aforesaid measures.

3.4.2 Data Collection from the FSS

Primary data were collected from the respondents of FSS (30 Key Informants). Frequent rapport and contact were maintained with the Key Informants. They were oriented towards the treatments conducted in the field in the post – scheduled sessions of Farmers and Scientists Results sharing meet.

3.4.3. Development of Semi – structured Interview Schedule

A semi structured interview schedule was prepared as a frame of reference for the investigation. The researcher exercised a greater degree of freedom in including or excluding certain questions and information according to the insight gained and from the vantage point of the respondents.

The semi structured interview schedule was divided into the following headings :

1. Degree of dependability in treatments in Banana
2. Status of adoption of treatments in Banana
3. Assignment of weightage score to treatments in Banana
4. Evaluative perception of treatments in Banana
5. Perception towards Participatory Technology Development (PTD)

3.4.4. Development of Treatment Rationalization Guide for ESS and RSS

The treatment rationalization guide was designed from the outcome of the two Farmers and Scientists Results sharing meets. The treatments were ranked based on the data presented in the meets and included in the final rationalization guide meant for the ESS and RSS.

3.4.5 Data Collection from the ESS and RSS

A Questionnaire was prepared by the researcher to assess the treatments in the following sub headings:

1. Degree of dependability in treatments in Banana
2. Status of adoption of treatments in Banana
3. Assignment of weightage score to treatments in Banana
4. Evaluative perception of treatments in Banana
5. Rationalization of Treatments in Banana
6. Perception towards Participatory Technology Development (PTD)

3.4.6. Documentation of Feed back

The experiences of experimented farmers were recorded both in the audio and visual form. The same has been incorporated in this report under result and discussion chapter.

3.5. OPERATIONALIZATION AND MEASUREMENT OF ATTRIBUTE MEASURES OF THE TREATMENTS IN BANANA

3.5.1. Degree of Dependability in Treatments of Banana

The degree of dependability in treatments of Banana was measured using an arbitrary scale of three point continuum with scores of 3,2,1 for 'Dependable', 'Somewhat dependable' and 'No dependable at all' respectively. All the three FSS, RSS and ESS were asked to express their opinion towards each treatment by choosing a score according to their dependability. Thus the maximum and minimum possible scores, for `dependability in treatments in Banana' were 36 and 12 respectively.

Treatment number	Treatment details	Degree of dependability		
		Dependable (3)	'Somewhat' Dependable (2)	No Dependable at all (1)
T_1				
T_2				
T_3				
T_4				
T_5				
T_6				
T_7				
T_8				
T_9				
T_{10}				
T_{11}				
T_{12}				

3.5.2. Status of Adoption of Treatments in Banana

The adoption of treatments in banana was measured using a two – point arbitrary scale of 'Done' (have done the treatments before) and 'Will Do' (will do the treatments in future) which had to be expressed on a dichotomy as 'Yes' or 'No'. All the three FSS, RSS and ESS are expressed their opinion by choosing either 'Yes' or 'No' option with a score of 4 and 1 for 'done' and 2 and 1 for 'Will Do'. The maximum score for each treatment was 48 for 'Done' and 24 for 'Will do' and a minimum score of 12 for both 'Done' and 'Will Do' situations.

Treatment number	Treatment details	Done		Will Do	
		Yes (4)	No (1)	Yes (2)	No (1)
T_1					
T_2					
T_3					
T_4					
T_5					
T_6					
T_7					
T_8					
T_9					
T_{10}					
T_{11}					
T_{12}					

3.5.3. Assignment Weightage Score to Treatments in Banana

FSS, RSS and ESS systems were facilitated to assign marks (out of a maximum of 10) to each treatment separately according to their perception and experiences.

Treatment number	Treatment details	Marks out of 10
T_1		
T_2		
T_3		
T_4		
T_5		
T_6		
T_7		
T_8		
T_9		
T_{10}		
T_{11}		
T_{12}		

3.5.4. Evaluative Perception of Treatments in Banana

The 'perceived effect' of treatment was measured using a three – point continuum (1 to 3) ranging from 'less effective, to 'most effective'. All the three FSS, RSS and ESS are expressed their opinion in the aforesaid continuum based upon their experience and perception.

Treatment number	Treatment details	Perceived Effect		
		Most effective (3)	Effective (2)	Less effective (1)
T_1				
T_2				
T_3				
T_4				
T_5				
T_6				
T_7				
T_8				
T_9				
T_{10}				
T_{11}				
T_{12}				

3.5.5. Rationalization of Treatments in Banana

The scale developed by Padaria and Singh (1990) to find out the scientific rationale of indigenous practices in dry farming was used in the present study with slight modification. The scale consisted of a three - point continuum viz., highly rational, rational, irrational with scores of 3, 2, and 1 respectively. All the three FSS, RSS and ESS were expressed their rationalization towards the treatments in the aforesaid continuum.

Treatment number	Treatment details	Rationalization		
		highly rational (3)	Rational (2)	Irrational (1)
T_1				
T_2				
T_3				
T_4				
T_5				

T_6
T_7
T_8
T_9
T_{10}
T_{11}
T_{12}

3.5.6. Perception towards Participatory Technology Development (PTD)

The Perception towards Participatory Technology Development (PTD) for RSS and ESS recorded together with the 24 statements relevant to the PTD in the researchers and Extensionists point of view. The respondents were asked give their responses in a three - point continuum of strongly agree, agree and disagree with scores of 3, 2, and 1 respectively. Similarly the Perception towards Participatory Technology Development (PTD) for FSS were elicited for 32 statements pertinent to the PTD in the farmers point of view and the respondents were asked to give their responses in a three - point continuum of strongly agree, agree and disagree with scores of 3,2, and 1 respectively.

3.6. STATISTICAL TOOLS USED IN THE STUDY

3.6.1 Kendall's Coefficient of Concordance (Kc)

Kendall's coefficient of concordance (Kc) measures the extent or degree of agreement among the respondents on a particular case. A significant Kc means that there is sufficient degree of agreement among the respondents with respect to the treatments on selected attributes.

In the present study, Kendall's coefficient of concordance (Kc) was used to find out the degree of agreement of FSS, RSS and ESS in their 'Degree of dependability in treatments in Banana', 'Status of adoption of treatments in Banana', 'Assignment of weightage score to treatments in Banana', and 'Evaluative perception of treatments in Banana'. The same statistical tool was used to compute the degree of agreement among the respondents from ESS and RSS on measures 'Scientific Rationality of Treatments in Banana' separately. The mean scores obtained were arranged in the descending order in such a way that the treatments with higher scores could be reckoned as better than the others.

The formula was,

$$Kc = \frac{\Sigma c_j^2 - (\Sigma ic_j)^2 / n}{1/12\, K^2\; n\,(n^2 - 1)}$$

Where,

c_j is the total of j th column for j = 1,2,........n

n is the number of observations in each variate

k is the number of variates

3.6.2. Spearman's Rank Order Correlation

Spearman's rank order correlation coefficient (r_s) was worked out to measure the relationship between the ranks of items under the study. It estimates the correlation between two characters on the basis of ranks of individuals in the whole lot for each of the characters without making an exact measurement for any of the individual.

In this study the spearman's rank order correlation (r_s) was compared with ESS, RSS and FSS for Degree of dependability in treatments in Banana', 'Status of adoption of treatments in Banana', 'Assignment of weightage score to treatments in Banana', 'Evaluative perception of treatments in Banana' and 'Scientific Rationality of Treatments in Banana'.

$$r_s = 1 - \frac{\Sigma d^2}{n\,(n^2 - 1)}$$

Where,

r_s – Spearmans rank order correlations coefficient

d – Difference between the two sets of values

n – Number of items.

3.6.3. Factor Analysis

Factor Analysis was worked out to measure the perception towards Participatory Technology Development (PTD) by FSS, RSS and ESS under the study. Factor analysis has two general uses – exploratory and confirmatory. The main goal is data reduction, that is, to simplify some number of measurements by identifying a smaller number of underlying traits, attitudes, or beliefs.

Principles

The theoretical basis for factor analysis is that variables correlated because they share one or more common components. That is, correlations among variables are explained by underlying factors. Mathematically a one – factor model for three variables can be represented as follows (Vs are variables, Fs are factors, Es represent random error).

$$V_1 = L_1*F_1 + E_1$$

$$V_2 = L_2*F_1 + E_2$$

$$V_3 = L_3*F_1 + E_3$$

Each variable is composed of the common factor (F_1) multiplied by a loading coefficient (L1, L2, L3 – the lambdas) plus a random component. If the factor were directly measurable (which it isn't) this would amount to a simple regression equation. Since these equations cannot be solved as given (the Ls, Fs and Es are unknowns), factor analysis takes an indirect approach. If the equations above hold, then consider why variables V_1 and V_2 correlate. Each contains an error (the E_1s, assumed to be random or unique) component that cannot contribute to their correlation (errors are assumed to have 0 correlation). However, they share the factor F_1, so if they correlate their correlation should be related to L_1 and L_2 (the factor loadings). If this logic is applied to all the pair wise correlations, the loading coefficients can be estimated from the correlation data. Thus one factor might account for the correlations in a set of variables. If not, the equations can be easily generalized to accommodate additional factors. There are different approaches to fitting factors to a correlation matrix (least squares, generalized least squares, maximum likelihood etc.) which have given rise to a number of factor methods.

3.6.5. Multivariate Analysis of Variance (MANOVA)

Multivariate Analysis of Variance (MANOVA) is a generalization of analysis of variance that permits testing for mean differences on several dependent measures simultaneously. Multivariate Analysis of Variance (MANOVA) tests for differences on a set or vector of means. MANOVA is typically performed for one of two reasons: statistical power and control of false positive results (known as Type I error).

3.7. OPERATIONALISATION OF CONCEPTS AND DEFINITIONS

Indigenous Technical Knowledge

It refers to the local knowledge/age old practices generated by farmers through their own non – formal experimentation considering their agro climatic conditions. They are low cost and no cost in nature, simple to follow, eco – friendly, have easy accessibility and have been proven as sustainable.

Participatory Technology Development (PTD)

Participatory Technology Development (PTD) is a creative process of joint experimentation and research by farmers and development agents in discovering ways of improving farmers' livelihoods.

Farmer Sub System (FSS)

The key informants representing two areas of the Banana belt in which the experiments were conducted who formed the farmer respondents of the study constitute Farmer Sub System (FSS).

Research Sub System (RSS)

The scientists who worked in Banana from Tamil Nadu Agricultural University (TNAU) and Indian Council of Agricultural Research (ICAR) who formed the researcher respondents of the study constitute the Research Sub System (RSS).

Extension Sub System (ESS)

The Assistant Directors of Horticulture (ADHs) and Horticultural Officers (HOs) of the two areas in which experiments were conducted who formed the extensionists respondents of the study constitute the Extension Sub System (ESS).

Relative Advantage

It is the degree to which an innovation or practice is perceived as being better that the idea it supersedes.

Compatibility

It is the degree to which an innovation is perceived as consistent with the existing values, past experiences, and needs of potential adopters. An innovation or practice can be compatible or incompatible with the socio – cultural values and beliefs, with previously introduced ideas or with client needs for innovations.

Complexity

It is the degree to which an innovation is perceived as relatively difficult to understand and use.

Simplicity

It is the degree to which an innovation is perceived as relatively easy to understand and use.

Trialability

It is the degree to which an innovation may be experimented with on a limited basis.

Observability

It is the degree to which the results of an innovation are visible to others.

Technology Attributes / Measurement Attributes

It denotes the various components of a technology deals with the complexity, compatibility, simplicity, trialability and so on.

Perceived Effect

This refers to ones perception or attitude towards a particular innovation or technology on various attributes of the technology.

Scientific Rationality

Rationality is the use of most effective means to reach a goal, which is not easily measurable. In the study, the scientific rationality means the scientific basis of a particular technology or treatment.

RESULTS AND DISCUSSION

The results of the research are presented and discussed under the following subheads:

4.1. Results of the experiments conducted in farmers' fields

4.2. Results of Path Analysis for the experiments conducted in farmers field

4.3. Benefit Cost Analysis of the selected treatments in banana

4.4. Attribute measures of the treatments in banana

4.5. Comparative analysis of attribute measures of the treatments in banana among Research Support System (RSS), Farmers Support System (FSS) and Extension Support System (ESS)

4.6. Perception of scientists, extensionists and farmers towards Participatory Technology Development (PTD)

4.7. Factor loadings of perception statements as perceived by scientists, extensionists and farmers

4.8. Multivariate Analysis of Variance (MANOVA) among three categories of respondents on Four Attributes

4.9. Results of the treatment combinations experienced by the practised farmers

4.10. The Researcher's self reflection on the Participatory Technology Development (PTD)

4.1. RESULTS OF THE EXPERIMENTS CONDUCTED IN FARMERS FIELDS

4.1.1. Results of the Experiments conducted at Kondayampatti Village with Grand – 9 Variety banana

The results of treatments conducted at Kondayamaptti village in Grand – 9 variety are depicted in Table 2. A perusal of the table reveals the selected yield parameters that effect the growth of Banana.

Table 2. Results of Field Experiments at Kondayampatti

Field – 1, 2, 3 Variety: GRAND -9

Treatments	Finger Length (cm)	Finger Girth (cm)	Finger Weight (gms)	Volume of finger (cc)	Density of fingers	No of fingers/ hand	Total Number of fingers/ bunch	No of hands/ bunch	Bunch Weight (Kg)	Yield t/ha	Yield increase (% over control)	TSS (⁰Brix)
T_1	21.00	12.67	59.02	252.00	1.26	25.33	136.00	16.00	29.51	59.02	101.00	15.17
T_2	21.67	12.33	58.38	252.67	1.28	26.33	139.00	16.67	29.19	58.38	99.93	15.17
T_3	21.33	12.00	56.99	254.00	1.25	25.00	137.00	16.00	28.50	56.99	97.55	15.00
T_4	21.00	13.00	61.00	251.67	1.27	26.00	138.00	16.67	30.50	61.00	104.42	15.00
T_5	19.00	10.67	53.30	251.67	1.25	22.33	130.00	15.00	26.65	53.30	91.24	14.83
T_6	19.33	11.33	59.13	252.33	1.27	24.67	135.00	15.67	29.57	59.13	101.21	15.17
T_7	19.67	11.33	64.12	255.67	1.28	29.00	140.00	17.33	32.06	64.12	109.76	14.83
T_8	24.67	14.33	72.00	261.33	1.31	32.33	150.00	18.67	36.00	72.00	123.24	15.67
T_9	23.00	13.33	55.62	253.33	1.21	26.00	135.00	15.00	27.81	55.62	95.21	15.17
T_{10}	24.00	14.00	69.38	260.00	1.30	31.67	147.00	18.33	34.69	69.38	118.76	15.50
T_{11}	21.67	12.00	61.20	254.67	1.28	26.33	136.00	15.67	30.60	61.20	104.76	15.00
T_{12}	17.67	10.00	58.42	242,00	1.05	15.33	115.00	11.33	29.21	58.42	100.00	14.17
S.Ed.	0.890	0.515	2.545	10.649	0.052	1.087	5.723	0.671	1.272	2.545		0.632
CD ((P=0.05)	1.847	1.068	5.279	22.084	0.108	2.255	11.87	1.39	2.64	5.279		1.312

4.1.1.1. Average Length of Finger

The average length of fingers was significantly increased by treatments as seen in Table 2. The application of fresh cow dung extracts and zinc sulphate in T8 (24.67) recorded maximum length of fruit followed by the application of urea, sulphate of potash, water and fresh cow dung in T10 (24.00) and were on par with each other. The control T12 recorded minimum length of fruit (17.67) which was on par with T7 (19.67) and T6 (19.33).

4.1.1.2. Average Girth of Finger

The average girth of fingers was significantly increased by treatments. The application of fresh cow dung extracts and zinc sulphate in T8 (14.33) recorded maximum girth of fruit followed by application of urea, sulphate of potash, water and fresh cow dung in T10 (14.00) and were on par with each other. The control T12 recorded minimum girth of fruit (10.00) which was on par with T5 (10.67).

4.1.1.3. Average Weight of Finger

The size of the individual fruit of banana, as represented by the average weight of finger was significantly increased by treatments. The maximum fruit size was observed due to the application of fresh cow dung extracts and zinc sulphate in T8 (72.00) followed by the application of urea, sulphate of potash, water and fresh cow dung in T10 (69.38) and were on par with each other. The minimum size of the fruit (58.42) was recorded in the untreated plantain.

4.1.1.4. Finger's Volume

The average volume of fingers was not significantly influenced by the treatments. However, the T8 recorded relatively higher volume (261.33) than the other treatments.

4.1.1.5. Density of Fruits

The average density of the fingers representing the filling capacity of the individual fruit, was found to be significantly higher in all the treatments than the control. Here the treatments from T1 to T11 were found to be on par and significantly higher than the control.

4.1.1.6. Number of fingers/hand

The data on the number of finger's per hand revealed that the highest number of fingers per hand (32.33) was recorded in the treatment T8 followed by T10 (31.67) and were on par. These two treatments were found to be significantly superior to the other treatments. The untreated (Control) T12 (15.33) banana recorded the lowest number of fingers per hand.

4.1.1.7. Total Number of fingers/bunch

The total number of fingers per bunch was significantly increased by treatments. The application of fresh cow dung extracts and zinc sulphate in T8 (150.00) recorded maximum number of fingers per bunch followed by the application of urea, sulphate of potash, water and fresh cow dung in T10 (147.00) and were on par with each other. The control T12 recorded minimum number of fingers per bunch (115.00).

4.1.1.8. Number of hands/bunch

The most important yield component of the Banana, the hands per bunch was significantly altered by various treatments. The spraying of fresh cow dung extracts and zinc sulphate in T8 (18.67) recorded the highest number of hands per bunch followed by spraying of urea, sulphate of potash, water and fresh cow dung in T10 (18.33) and were on par with each other. The lowest (11.33) number of hands per bunch was observed in control.

4.1.1.9. Bunch Weight

The bunch weight was significantly influenced by treatments. The highest bunch weight (36.00) was observed in banana due to the application of fresh cow dung extracts and zinc sulphate in (T8) followed by the application of urea, sulphate of potash, water and fresh cow dung in T10 (34.69 kg/bunch) which were on par with each other. The untreated plantain (Control) recorded the lowest bunch weight (29.21 kg/bunch).

4.1.1.10. Yield

The treatments had a significant influence on the yield of Banana and the highest yield was recorded in T8 (72.00 t/ha) followed by T10 (69.38 t/ha) which were on par and were significantly superior to the rest of the treatments. There was an increase of 23.24 per cent and 18.76 per cent over the control in treatments T8 and T10 respectively. The lowest yield (58.42 t/ha) was recorded in T12.

4.1.1.11. TSS

The data on the TSS revealed that the highest TSS (15.67) was recorded in the treatment T8 followed by T10 (15.50) and were on par. These two treatments were found to be significantly superior to the other treatments. The untreated (Control) T12 (14.17) plantain recorded the lowest number of fingers per hand.

4.1.2. Results of the Experiments conducted at K. Dharmathupatti Village in Robusta Variety

The results of treatments conducted at K. Dharmathupatti village in Robusta variety are depicted in Table 3. A scrutiny of the data reveals the yield parameters that effect the yield of Banana.

Table 3. Results of Field Experiment at K. Dharmathupatti - Robusta

Field – 1 Variety: Robusta

Treatments	Finger Length (cm)	Finger Girth (cm)	Finger Weight (gms)	Volume of finger (cc)	Density of fingers	No of fingers / hand	Total Number of fingers/ bunch	No of hands/ bunch	Bunch Weight (Kg)	Yield t/ha	Yield increase (% over control)	TSS (⁰ Brix)
T_1	21.00	12.00	270.0	225.00	1.24	24.00	118.00	10.00	31.9	57.35	133.18	13.5
T_2	22.00	14.00	262.0	221.00	1.23	20.00	114.00	8.00	29.9	53.76	124.84	13.5
T_3	21.00	13.00	270.0	224.00	1.25	22.00	118.00	10.00	31.9	57.35	133.18	13.5
T_4	23.00	14.00	262.0	220.00	1.24	22.00	105.00	8.00	27.5	49.52	115.00	13.0
T_5	20.00	12.00	260.0	218.00	1.24	18.00	112.00	7.00	29.1	52.42	121.73	13.0
T_6	20.00	12.00	263.0	220.00	1.24	20.00	116.00	8.00	30.5	54.91	127.51	13.5
T_7	20.00	12.00	265.0	221.00	1.24	21.00	117.00	9.00	31.0	55.81	129.60	13.5
T_8	22.00	13.00	273.0	228.00	1.24	26.00	123.00	12.00	33.6	60.44	140.36	14.5
T_9	21.00	13.00	271.0	226.00	1.24	25.00	120.00	11.00	32.5	58.54	135.94	14.0
T_{10}	23.00	15.00	268.0	223.00	1.25	24.00	117.00	10.00	31.4	56.44	131.07	13.5
T_{11}	20.00	12.00	266.0	222.00	1.24	22.00	116.00	9.00	30.9	55.54	128.98	13.5
T_{12}	18.00	10.00	230.0	202.00	1.19	15.00	104.00	6.00	23.9	43.06	100.00	12.0
SEd	0.877	0.532	11.06	9.281	0.052	0.912	4.841	0.384	1.278	2.30		0.564
CD (P=.05)	1.818	1.103	22.95	19.24	0.107	1.89	10.04	0.797	2.65	4.771		1.170

4.1.2.1. Average Length of Finger

The average length of fingers was considerably increased by treatments. The application of *panchakavia* and household wood ash in T4 (23.00) and application of urea, sulphate of potash, water and fresh cow dung in T10 (23.00) recorded maximum length of fruit followed by application of brick chamber wood ash and pregnant cow urine in T2 (22.00) and application of fresh cow dung extracts and zinc sulphate in T8 (22.00) and were on par with each other. The control recorded minimum length of fruit (18.00) which was on par with T6 (20.00) T7 (20.00) and T11 (20.00).

4.1.2.2. Average Girth of Finger

The average girth of fingers was significantly increased by various treatments. The application of urea, sulphate of potash, water and fresh cow dung in T10 (15.00) recorded maximum girth of fruit followed by application of brick chamber wood ash and pregnant cow urine in T2 (14.00) and application of *panchakavia* and household wood ash in T4 (14.00) and were on par with each other. The control T12 recorded minimum girth of fruit (10.00).

4.1.2.3. Average Weight of Finger

The size of the individual fruit of banana, as represented by the average weight of finger was not affected by the treatments. The maximum fruit size was observed due to the application of fresh cow dung extracts and zinc sulphate in T8 (273.00) followed by the application of fresh cow dung extracts and borax in T9 (271.00) and were on par with each other. The minimum size of the fruit (230.00) was recorded in the untreated plantain.

4.1.2.4. Finger's Volume

The average volume of fingers was not affected by the treatments. However, the T8 recorded comparatively higher volume (228.00) than the other treatments.

4.1.2.5. Density of Fruits

The average density of the fingers representing the filling capacity of the individual fruit, was not affected by the treatments. Here the treatments from T1 to T11 were found to be on par and considerably higher than the control.

4.1.2.6. Number of fingers/hand

The data on the number of finger's per hand revealed that the highest number of fingers per hand (26.00) was recorded in the treatment T8 followed by T9 (25.00) and were on par. These two treatments were found to be significantly superior to the other treatments. The untreated (Control) T12 (15.00) plantain recorded the lowest number of fingers per hand.

4.1.2.7. Total Number of fingers/bunch

The total number of fingers per bunch was significantly increased by treatments. The application of fresh cow dung extracts and zinc sulphate in T8 (123.00) recorded maximum number of fingers per bunch followed by application of fresh cow dung extracts and borax in T9 (120.00) and were on par with each other. The control T12 recorded minimum number of fingers per bunch (104.00) which was on par with T4 (105.00)

4.1.2.8. Number of hands/bunch

The most important yield component of the Banana, the hands per bunch was significantly changed by various treatments. The spraying of fresh cow dung extracts and zinc sulphate in T8 (12.00) recorded the highest number of hands per bunch followed by spraying of fresh cow dung extracts and borax in T9 (11.00) and were on par with each other. The lowest (6.00) number of hands per bunch was seen in control.

4.1.2.9. Bunch Weight

The bunch weight was significantly influenced by treatments. The highest bunch weight (33.60) was observed in plantain due to the application of fresh cow dung extracts and zinc sulphate in T8 followed by the application of fresh cow dung extracts and borax in T9 (32.5 kg/bunch) which were on par with each other. The untreated plantain (Control) recorded the lowest bunch weight (23.90 kg/bunch).

4.1.2.10. Yield

The treatments had a considerable influence on the yield of Banana. The highest yield was recorded in T8 (60.44 t/ha) followed by T9 (58.54 t/ha) which were on par and were significantly superior to the rest of the treatments. It was found to be an increase of 40.36% and 35.94% over the control in treatments T8 and T9 respectively. The lowest yield (43.06 t/ha) was recorded in T12.

4.1.2.11. TSS

The data on the TSS revealed that the highest TSS (14.50) was recorded in the treatment T8 followed by T9 (14.00) and were on par. These two treatments were found to be significantly superior to the other treatments. The untreated (Control) T12 (12.00) plantain recorded the lowest TSS

4.1.3. Results of the Experiments conducted at K. Dharmathupatti Village in Poovan Variety

The results of treatments conducted at K. Dharmathupatti village in Poovan variety are specified in Table 4. An analysis of the table reveals the following yield parameters that effect the yield of Banana.

Table 4. Results of Field Experiments at K. Dharmathupatti - Poovan

Field – 2 Variety: Poovan

Treatments	Finger Length (cm)	Finger Girth (cm)	Finger Weight (gms)	Volume of finger (cc)	Density of fingers	No of fingers/ hand	Total Number of fingers/ bunch	No of hands/ bunch	Bunch Weight (Kg)	Yield t/ha	Yield increased (% over control)	TSS (0 Brix)
T_1	13.00	12.00	120.00	76.00	1.45	15.00	158.00	10.00	19.0	35.55	147.38	14.5
T_2	12.00	11.00	116.00	77.00	1.38	14.00	145.00	9.00	16.8	31.54	130.76	14.5
T_3	14.00	12.00	118.00	76.00	1.42	16.00	156.00	11.00	18.4	34.52	143.11	14.5
T_4	14.00	13.00	105.00	75.00	1.27	12.00	140.00	9.00	14.7	27.56	114.26	14.5
T_5	13.00	12.00	109.00	74.00	1.34	11.00	146.00	8.00	15.9	29.84	123.71	14.5
T_6	12.00	11.00	108.00	76.00	1.29	15.00	158.00	10.00	17.1	32.00	132.66	14.5
T_7	12.00	11.00	100.00	76.00	1.18	12.00	163.00	8.00	16.3	30.56	126.69	13.5
T_8	15.00	13.00	120.00	75.00	1.47	16.00	160.00	10.00	19.2	36.00	149.25	14.5
T_9	13.00	12.00	130.00	75.00	1.60	15.00	160.00	10.00	20.8	39.00	161.69	14.5
T_{10}	14.00	13.00	110.00	77.00	1.30	13.00	150.00	9.00	16.5	30.94	128.27	14.0
T_{11}	12.00	11.00	115.00	76.00	1.38	15.00	150.00	10.00	17.3	32.34	134.07	14.5
T_{12}	11.00	10.00	96.00	71.00	1.21	10.00	134.00	7.00	12.9	24.12	100.00	13.0
SEd	0.543	0.493	4.746	3.165	0.057	0.583	6.393	0.393	0.725	1.360		0.599
CD (P=.05)	1.127	1.022	9.843	6.56	0.119	1.209	13.25	0.816	1.504	2.82		1.243

4.1.3.1. Average Length of Finger

The average length of fingers was significantly increased by treatments. The application of fresh cow dung extracts and zinc sulphate in T8 (15.00) recorded maximum length of fruit followed by application of urea, sulphate of potash, water and fresh cow dung in T10 (14.00), panchakavia and brick chamber wood ash in T3 (14.00) and application of panchakavia and household wood ash in T4 (14.00) and were on par with each other. The control recorded minimum length of fruit (11.00) which was on par with T2 (12.00), T6 (12.00), T7 (12.00) and T11 (12.00) respectively.

4.1.3.2. Average Girth of Finger

The average girth of fingers was significantly increased by treatments. The application of panchakavia and household wood ash in T4 (13.00), application of fresh cow dung extracts and zinc sulphate in T8 (13.00) and the application of urea, sulphate of potash, water and fresh cow dung in T10 (13.00) recorded maximum girth of fruit and were on par with each other. The control T12 recorded minimum (10.00) girth of fruit.

4.1.3.3. Average Weight of Finger

The size of the individual fruit of banana, represented by the average weight of finger was significantly increased by various treatments. The maximum fruit size was gained due to the application of fresh cow dung extracts and borax in T9 (130.00) followed by the application household wood ash and pregnant cow urine in T1 (120.00) application of fresh cow dung extracts and zinc sulphate in T8 (120.00) and were on par with each other. The minimum size of the fruit (96.00) was traced in the untreated plantain.

4.1.3.4. Finger's Volume

The average volume of fingers was not influenced by the treatments. However, the T2 and T10 witnessed relatively higher volume (77.00) than the other treatments.

4.1.3.5. Density of fruits

The average density of the fingers representing the filling capacity of the individual fruit, was significantly increased by treatments. The maximum density was obtained due to the application of fresh cow dung extracts and borax in T9 (1.60) and followed by application of fresh cow dung extracts and zinc sulphate in T8 (1.47) and were on par with each other. The control T12 recorded minimum density of fruit (1.21) which was on par with T24 (1.27) and T6 (1.29).

4.1.3.6. Number of fingers/hand

The data on the number of fingers per hand showed that the highest number of fingers per hand (16.00) was recorded in the treatment T8 and T3 followed by (15) in T1, T6, T9, and T11 and were on par. These treatments were found to be significantly superior to the other treatments. The untreated (Control) T12 (10.00) plantain confirmed the lowest number of fingers per hand.

4.1.3.7. Total Number of fingers/bunch

The total number of fingers per bunch was significantly increased by treatments. The application of fresh cow dung extracts, zinc sulphate and borax in T7 recorded maximum number of fingers per bunch (163.00) followed by application of fresh cow dung extracts and zinc sulphate in T8 (160.00), application of fresh cow dung extracts and borax in T9 (160.00) and were on par with each other. The control T12 recorded minimum number of fingers per bunch (134.00) which was on par with T4 (140.00).

4.1.3.8. Number of hands/bunch

The most important yield component of the Banana, the hands per bunch was significantly altered by differential treatments. The spraying of panchakavia and brick chamber wood ash in T3 (11.00) recorded the highest number of hands per bunch followed by T1, T6, T8, T9 and T11 (10.00) and were on par with each other. The lowest number of hands per bunch (7.00) was recorded in control.

4.1.3.9. Bunch Weight

The bunch weight was significantly influenced by the treatments. The highest (20.80) bunch weight was observed in plantain due to the application of fresh cow dung extracts and borax in T9 followed by the application of fresh cow dung extracts and zinc sulphate in T8 (19.20 kg/bunch) which were on par with each other. The untreated plantain (Control) recorded the lowest (12.90 kg/bunch) bunch weight.

4.1.3.10. Yield

The treatments had a significant influence on the yield of Banana. The highest yield was recorded in T9 (39.00 t/ha) followed by T8 (36.00 t/ha) which were on par and were significantly superior to the rest of the treatments. It was found to be an increase of 61.69 per cent and 49.25 per cent over the control in treatments T9 and T8 respectively. The lowest yield (24.12 t/ha) was recorded in T12.

4.1.3.11. TSS

The data on the TSS revealed that the highest TSS (14.50) was recorded in all the treatment except T7 and were on par. These treatments were not influenced by the yield. The untreated (Control) T12 (13.00) plantain recorded the lowest TSS.

4.1.4. Results of the Experiments conducted at K. Dharmathupatti Village in Red Banana Variety

The results of treatments accomplished at K. Dharmathupatti village in Red Banana variety arc illustrated in Table 5. An inspection of the table reveals the following yield parameters that affect the yield of Banana.

Table 5. Results of Field Experiments at K. Dharmathupatti – Red Banana

Field – 3 Variety: Red Banana

Treat-ments	Finger Length (cm)	Fin-ger Girth (cm)	Fin-ger Weight (gms)	Volu-me of finger (cc)	Den-sity of fingers	No of fing-ers/ hand	Total Num-ber of fingers/ bunch	No of ha-nds/ bunch	Bunch Wei-ght (Kg)	Yield t/ha	Yield incre-ased (% over control)	TSS (° Brix)
T_1	19.00	12.00	299.00	118.00	2.53	11.00	97.00	6.00	32.0	52.21	117.72	12.00
T_2	20.00	14.00	300.00	120.00	2.50	12.00	100.00	7.00	33.0	54.00	121.75	12.00
T_3	20.00	15.00	298.00	120.00	2.48	11.00	98.00	6.00	32.2	52.57	118.53	12.00
T_4	19.00	14.00	305.00	122.00	2.50	13.00	101.00	8.00	33.9	55.45	125.02	12.50
T_5	18.00	12.00	293.00	118.00	2.48	10.00	96.00	5.00	31.1	50.63	114.16	12.00
T_6	18.00	13.00	293.00	117.00	2.50	10.00	97.00	5.00	31.4	51.16	115.35	12.50
T_7	18.00	12.00	297.00	119.00	2.50	11.00	107.00	6.00	34.7	57.20	128.97	12.50
T_8	20.00	15.00	370.00	129.00	3.64	18.00	110.00	10.00	44.4	73.26	165.18	13.50
T_9	18.00	12.00	350.00	125.00	3.60	15.00	110.00	9.00	42.0	69.30	156.25	12.50
T_{10}	21.00	15.00	365.00	128.00	3.63	17.00	106.00	10.00	42.3	69.64	157.02	13.00
T_{11}	18.00	12.00	300.00	125.00	3.20	13.00	95.00	8.00	31.5	51.30	115.67	11.50
T_{12}	11.00	10.00	280.00	108.00	2.59	9.00	88.00	5.00	27.4	44.35	100.00	11.00
SEd	0.772	0.548	13.16	5.069	0.121	0.532	4.22	0.303	1.467	2.405	0.514	
CD (P=.05)	1.601	1.138	27.31	10.51	0.25	1.103	8.75	0.629	3.04	4.98	1.067	

4.1.4.1. Average Length of Finger

The average length of fingers was significantly increased by treatments. The application of urea, sulphate of potash, water and fresh cow dung in T10 (21.00) recorded maximum length of fruit followed by application of brick chamber wood ash and pregnant cow urine in T2 (20.00), application panchakavia and brick chamber wood ash in T3 (20.00) and application of fresh cow dung extracts and zinc sulphate in T8 (20.00) and were on par with each other. The control recorded minimum (11.00) length of fruit.

4.1.4.2. Average Girth of Finger

The average girth of fingers was significantly increased by treatments. The application of panchakavia and brick chamber wood ash in T3 (15.00), application of fresh cow dung extracts and zinc sulphate in T8 (15.00) and application of urea, sulphate of potash, water and fresh cow dung in T10 (15.00) recorded maximum girth of fruit and were on par with each other. The control T12 recorded minimum (10.00) girth of fruit.

4.1.4.3. Average Weight of Finger

The size of the individual fruit of banana, as represented by the average weight of finger had significantly increased by treatments. The maximum fruit size (370.00) was observed due to the application of fresh cow dung extracts and zinc sulphate in T8 followed by the application of urea, sulphate of potash, water and fresh cow dung in T10 (365.00) and were on par with each other. The minimum size of the fruit (280.00) was recorded in the untreated plantain in T12.

4.1.4.4. Finger's Volume

The average volume of fingers had significantly increased by treatments. The maximum fingers volume was observed due to the application of fresh cow dung extracts and zinc sulphate in T8 (129.00) followed by the application of urea, sulphate of potash, water and fresh cow dung in T10 (128.00) and were on par with each other. The minimum size of the fruit (108.00) was recorded in the untreated plantain in T12.

4.1.4.5. Density of Fingers

The average density of the fingers representing the filling capacity of the individual fruit was found to be significantly higher in all the treatments than the control. The maximum average density of the fingers was observed due to the application of fresh cow dung extracts and zinc sulphate in T8 (3.64) followed by the application of urea, sulphate of potash, water and fresh cow dung in T10 (3.63) and were on par with each other. The minimum size of the fruit (2.59) was recorded in the untreated plantain in T12.

4.1.4.6. Number of fingers/hand

The data on the number of fingers per hand revealed that the highest number of fingers per hand (18.00) was recorded in the treatment T8 followed by T10 (17.00) and were on par. These two treatments were found to be significantly better to the other treatments. The untreated (Control) in T12 (9.00) plantain recorded the lowest number of fingers per hand.

4.1.4.7. Total Number of fingers/bunch

The total number of fingers per bunch was significantly increased by treatments. The application of fresh cow dung extracts and zinc sulphate in T8 (110.00), application of fresh cow dung extracts and borax in T9 (110.00) and were on par with each other. The control T12 recorded minimum number of fingers per bunch (88.00).

4.1.4.8. Number of hands/bunch

The most important yield component of the Banana, the hands per bunch was significantly altered by various treatments. The spraying of application of fresh cow dung extracts and zinc sulphate in T8 (10.00) and the application of urea, sulphate of potash, water and fresh cow dung in T10 (10.00) recorded the highest number of hands per bunch and were on par with each other. The lowest number of hands per bunch was observed in control.

4.1.4.9. Bunch Weight

The bunch weight was significantly influenced by treatments. The highest bunch weight (44.4) was observed in plantain due to the application of fresh cow dung extracts and zinc sulphate in T8 followed by the application of urea, sulphate of potash, water and fresh cow dung in T10 (42.30 kg/bunch) which were on par with each other. The untreated plantain (Control) recorded the lowest bunch weight (27.40 kg/bunch).

4.1.4.10. Yield

The treatments had a significant influence on the yield of Banana. The highest yield (73.26 t/ha) was recorded in T8 followed by T10 (69.64 t/ha) which were on par and were significantly superior to the rest of the treatments. It was found to be an increase of 65.18 per cent and 57.02 per cent over the control in treatments T8 and T10 respectively. The lowest yield (44.35 t/ha) was recorded in T12.

4.1.4.11. TSS

The data on the TSS revealed that the highest TSS (13.50) was recorded in the treatment T8 followed by T10 (13.00) and were on par. These two treatments were found to be significantly superior to the other treatments. The untreated (Control) T12 plantain recorded the lowest TSS (11.00).

Discussion

The results obtained in the present study are discussed with the support of relevant literature in this season. The yield of banana in all the four varieties namely Grand 9, Robusta, Poovan and Red banana were significantly increased by the treatments.

The application of fresh cow dung extract and zinc sulphate in treatment (T_8) significantly increased the yield of banana in all the varieties except the poovan variety where the highest yield was obtained in treatment (T_9) by application of fresh cow dung extract and borax. This was due to the significantly increased bunch weight. The bunch weight also found to follow the trend similar to that of the yield of Banana (t/ha). Yield in Banana is a function of bunch weight and number of plants per hectare. (Kumar & Kumar, 2008).

The higher yield of banana obtained in the elite treatment T8 might be due to more number of fingers/bunch, fingers/hand, hands/bunch, average finger weight, finger volume etc., Generally, the farmer practice is to cut the distal end of the inflorescence after the opening of the last hand. Due to the inadequate supply and flow of nutrients and hormones the distal hand in bunch may not be fully developed into fruit. In the present study in treatment T8 by application of fresh cow dung extract with zinc sulphate an increased the number of fingers/bunch. The presence of nutrients in the available form as well as plant growth hormones (Adegunloye, 2007) might have been responsible for the retention of the fingers in the last hands of the bunch.

Zinc is required for the synthesize of Auxin, a hormone responsible for increasing the fruit set. In the present study, the inclusion of zinc as one of the ingredients in T8 treatment also might have played an important role in increasing the number of fingers per bunch.

The number of fingers per hand, another important yield component closely follows the trend similar to that of total number of fingers per bunch. In fact the number of fingers per hand is reflected in total number of fingers per bunch. In the present study the T8 consistently registered more number of fingers per hand in all the four varieties.

In banana, the size of the finger attracts the consumer and fetches higher price. Any technology enhancing this important yield component is worth rewarding and will attract the farmers to adopt. In the present study the maximum weight of fingers had registered in T8 except poovan; In Poovan the highest weight was registered due to T_9 by application of fresh cow dung extract and borax. The weight of the bunch in poovan was also found to be high in T_9. The differential response of banana cultivars to treatments has been reported by Kotur and Keshava Murthy, 2007. The maximum weight of fingers in T8 due to the higher mobilization of starch produced during photosynthesis from the leaves. (Kumar and Jeyakumar, 2001) and the above process was accelerated by the plant hormones present in the cow dung. This is also evident from the analytical report of cow dung, cow urine, panchakavia etc. The analytical data is furnished in the appendix - 4. Prasad and Power (1997) reported that the fresh cow dung is rich in plant growth hormones as the cow grazes the tip of the grasses. The tips of the grasses are usually rich in hormones viz., auxin, cytokinin and gibberlic acid.

The average finger weight was also reflected in the finger volume. As the size increases the volume also increased. The enhanced girth of fingers was also found to follow the trend similar to that of finger weight. The finger length, the other finger components differ widely due to treatment. Hence, the size of the fingers as measured by weight was due to increased girth rather than due to length. The hormones available in active form in cow dung as well as in inorganic zinc which is indispensable for the synthesize of IAA could have increased the cell size, which may be the probable reason for the increased girth of fruits.

TSS is the measurement of sugars which is responsible for the sweet taste in fruit, decides the quality of the fruit. In the present study, the highest TSS was registered in T8 in all the four varieties. This is in close conformity of the findings of Kumar and Jeyakumar (2001). This might be probably due to the highest accumulation of starch and subsequently disintegrated into starch on ripening. (Prevel, 1989). The correlation coefficient between yield and yield components irrespective of varieties and treatments revealed that the bunch weight was significantly and positively associated with number of fingers per bunch, volume of fingers and fingers weight.

Hence, it could be concluded from the present investigation that the application of fresh cow dung extract and zinc sulphate (T8) to the cut distal end of inflorescence (male axis). The distal end of inflorescence leaving 15 cm was cut, 15 days after opening of the last hand enhances the yield of banana in cultivars Grand 9, Robusta, Poovan and Red banana. The enhanced yield was due to increased number of fingers and weight of fingers. The application of fresh cow dung extract and borax (T9) resulted in higher yield of banana in Poovan.

4.2. RESULTS OF PATH ANALYSIS FOR THE EXPERIMENTS CONDUCTED IN FARMERS FIELD

4.2.1. Path Analysis – Simple Correlation Co-efficient for Grand - 9 variety at Kondayampatti village

Simple correlation Co-efficient between yield and yield components viz., Bunch weight, Number of hands per bunch, Number of fingers per hand, Average weight of finger, Volume of finger, Density of fingers, Finger length, Finger girth, TSS and Total Number of fingers per bunch in Grand - 9 variety at Kondayampatti village are presented in Table 6.

A perusal of the data indicate that the higher positive and significant association was obtained between yield and bunch weight and finger weight (r = +1.000), followed by number of fingers per hand (r = +0.720) suggesting that any increase in the above parameter could lead to enhanced yield of fruits in banana. Correlation between the yield component revealed that the bunch weight was highly associated with the finger weight (r = +1.000), number of fingers per hand (r = +0.720). It indicated that the bunch weight could be increased by increasing the size of the fingers (i.e. finger weight) and number of fingers.

Table 6. Results of Path analysis – Simple Correlation Co – efficient for Grand – 9 variety

Character	Yield	Bunch Weight	No of hands per bunch	No of fingers per hand	Finger Weight	Volume of finger	Density of fingers	Finger length	Finger girth	TSS	Total Number of fingers per bunch
Yield	1.000	1.000**	0.682*	0.720**	1.000**	0.688*	0.435	0.611*	0.622*	0.569	0.688*
Bunch Weight		1.000	0.682*	0.720**	1.000**	0.688*	0.435	0.611*	0.622*	0.569	0.688*
No of hands per bunch			1.000	0.968**	0.682*	0.932**	0.921**	0.730**	0.752**	0.859**	0.984**
No of fingers per hand				1.000	0.720**	0.969**	0.872**	0.821**	0.815**	0.883**	0.985**
Finger Weight					1.000	0.688*	0.435	0.611*	0.622*	0.569	0.688*
Volume of finger						1.000	0.861**	0.822**	0.768**	0.886**	0.960**
Density of fingers							1.000	0.595*	0.600*	0.819**	0.901**
Finger length								1.000	0.956**	0.870**	0.826**
Finger girth									1.000	0.876**	0.829**

TSS	1.000	0.918**
Total Number of fingers per bunch		1.000

** Significant at 1% level; Table value for 1% level: 0.708 (d.f) (n-2) = 10

* Significant at 5% level; Table value for 5% level: 0.576

4.2.2. Path Analysis – Simple Correlation Co-efficient for Robusta Variety in K. Dharmathupatti

Simple correlation Co-efficient between yield and yield components viz., Bunch weight, Number of hands per bunch, Number of fingers per hand, Average weight of finger, Volume of finger, Density of fingers, Finger length, Finger girth, TSS and Total Number of fingers per bunch in Robusta variety at K. Dharmathupatti village are presented in Table 7.

Table 7. Results of Path analysis – Simple Correlation Co – efficient for Robusta variety

Character	Yield	Bunch Weight	No of hands per bunch	No of fingers per hand	Finger Weight	Volume of finger	Density of fingers	Finger length	Finger girth	TSS	Total Number of fingers per bunch
Yield	1.000	1.000**	0.892**	0.851**	0.935**	0.942**	0.797**	0.448	0.443	0.749**	0.963**
Bunch Weight		1.000	0.892**	0.852**	0.935**	0.942**	0.797**	0.450	0.445	0.749**	0.963**
No of hands per bunch			1.000	0.949**	0.813**	0.851**	0.616*	0.517	0.450	0.893**	0.862**
No of fingers per hand				1.000	0.868**	0.900**	0.706*	0.685*	0.595*	0.867**	0.754**
Finger Weight					1.000	0.993**	0.929**	0.639*	0.615*	0.889**	0.806**
Volume of finger						1.000	0.882**	0.659*	0.619*	0.923**	0.821**
Density of fingers							1.000	0.634*	0.657*	0.701*	0.633*

Finger length	1.000	0.951**	0.519	0.258
Finger girth		1.000	0.487	0.273
TSS			1.000	0.907**
Total Number of fingers per bunch				1.000

** Significant at 1% level; Table value for 1% level: 0.708 (d.f) (n-2) = 10

* Significant at 5% level; Table value for 5% level: 0.576

A perusal of the data indicated that the higher positive and significant association was obtained between yield and bunch weight (r = +1.000), followed by the fingers per bunch (r = +0.963), volume of finger (r = +0.942), finger weight (r = +0.935) and number of hands per bunch (r = +0.892) suggesting that any increase in the above parameter could lead to enhanced yield of fruits in banana.

Correlation between the yield component reveal that the bunch weight was highly associated with the fingers per bunch (r = +0.963), volume of finger (r = +0.942), finger weight (r = +0.935). It indicated that the bunch weight could increased by increasing the number of fingers per bunch and size of the fingers (volume and finger weight).

4.2.3. Path Analysis – Simple Correlation Co-efficient for Poovan Variety in K. Dharmathupatti

Simple correlation Co-efficient between yield and yield components viz., Bunch weight, Number of hands per bunch, Number of fingers per hand, Average weight of finger, Volume of finger, Density of fingers, Finger length, Finger girth, TSS and Total Number of fingers per bunch in Poovan variety at K. Dharmathupatti village are presented in Table 8.

A perusal of the data indicated that the higher positive and significant association was obtained between yield and bunch weight (r = +1.000), followed by finger weight (r = +0.925), density of fingers (r = +0.876), number of fingers per hand (r = +0.848), number. of fingers per bunch (r = +0.801), and number of hands per bunch (r = + 0.789) and suggesting that any increase in the above parameter could lead to enhanced yield of fruits in banana.

Table 8. Results of Path analysis – Simple Correlation Co – efficient for Poovan variety

Character	Yield	Bunch Weight	No of hands per bunch	No of fingers per hand	Finger Weight	Volume of finger	Density of fingers	Finger length	Finger girth	TSS	Total Number of fingers per bunch
Yield	1.000	1.000**	0.789**	0.848**	0.925**	0.504	0.876**	0.439	0.374	0.630*	0.801**
Bunch Weight		1.000	0.791**	0.850**	0.924**	0.502	0.876**	0.434	0.370	0.629*	0.801**
No of hands per bunch			1.000	0.951**	0.768**	0.594*	0.687*	0.497	0.393	0.759**	0.584*
No of fingers per hand				1.000	0.794**	0.596*	0.713**	0.413	0.280	0.677*	0.673*
Finger Weight					1.000	0.421	0.978**	0.446	0.404	0.723**	0.515
Volume of finger						1.000	0.224	0.306	0.350	0.563	0.538
Density of fingers							1.000	0.421	0.368	0.657*	0.426
Finger length								1.000	0.950**	0.507	0.291
Finger girth									1.000	0.518	0.219
TSS										1.000	0.322
Total Number of fingers per bunch											1.000

** Significant at 1% level; Table value for 1% level: 0.708 (d.f) (n-2) = 10

* Significant at 5% level; Table value for 5% level: 0.576

Correlation between the yield component revealed that the bunch weight was highly associated with the finger weight (r = +0.924), density of fingers (r = +0.876), number of fingers per hand (r =

+0.850), number of fingers per bunch (r = +0.801) and number of hands per bunch (r = + 0.791). It indicated that the bunch weight could be increased by increasing the size of the fingers (volume and finger weight) and number of fingers per hand.

4.2.4. Path Analysis – Simple Correlation Co-efficient for Red Banana Variety in K. Dharmathupatti

Simple correlation Co-efficient between yield and yield components viz., Bunch weight, Number of hands per bunch, Number of fingers per hand, Average weight of finger, Volume of finger, Density of fingers, Finger length, Finger girth, TSS and Total Number of fingers per bunch in Red banana variety at K. Dharmathupatti village are presented in Table 9.

A perusal of the data indicated that the higher positive and significant association was obtained between yield and bunch weight (r = +1.000), followed by finger weight (r = +0.973), number of fingers per hand (r = +0.932), number of fingers per bunch (r = +0.918), number of hands per bunch (r = + 0.873) and density of fingers (r = +0.837) and suggesting that any increase in the above parameter could lead to enhanced yield of fruits in banana.

Table 9. Results of Path analysis – Simple Correlation Co – efficient for Red Banana variety

Character	Yield	Bunch Weight	No of hands per bunch	No of fingers per hand	Finger Weight	Volume of finger	Density of fingers	Finger length	Finger girth	TSS	Total Number of fingers per bunch
Yield	1.000	1.000**	0.873**	0.932**	0.973**	0.836**	0.837**	0.551	0.547	0.856**	0.918**
Bunch Weight		1.000	0.876**	0.935**	0.976**	0.838**	0.841**	0.551	0.550	0.855**	0.913**
No of hands per bunch			1.000	0.977**	0.903**	0.888**	0.875**	0.509	0.545	0.644*	0.720**
No of fingers per hand				1.000	0.960**	0.904**	0.890**	0.555	0.601*	0.754**	0.773**
Finger Weight					1.000	0.826**	0.903**	0.498	0.544	0.792**	0.804**

Volume of finger	1.000	0.758**	0.777**	0.673*	0.737**	0.770**
Density of fingers		1.000	0.257	0.271	0.520	0.612*
Finger length			1.000	0.832**	0.682*	0.614*
Finger girth				1.000	0.694*	0.512
TSS					1.000	0.865**
Total Number of fingers per bunch						1.000

** Significant at 1% level; Table value for 1% level: 0.708 (d.f) (n-2) = 10

* Significant at 5% level; Table value for 5% level: 0.576

Correlation between the yield component revealed that the bunch weight was highly associated with the finger weight (r = +0.976), number of fingers per hand (r = +0.935), number of fingers per bunch (r = +0.913), number of hands per bunch (r = + 0.876) and density of fingers (r = +0.841). It indicated that the bunch weight could increased by increasing the size of the fingers (finger weight), number of fingers per hand and number of fingers per bunch.

4.3. BENEFIT COST ANALYSIS OF THE SELECTED TREATMENTS

4.3.1. Benefit Cost Ratio of the treatments in Banana (Grand 9 Variety)

The analysis of Benefit Cost Ratio of the treatments in Banana (Grand 9 Variety) is presented in Table. 10

Table 10. Benefit Cost Ratio of the treatments in Banana (Grand 9 Variety)

Treatments	Yield (t/ha)	Gross Income	Cost of cultivation	Net income	Benefit cost ratio
T_1	59.02	2,36,096	98,250	1,37,846	2.4
T_2	58.38	2,33,520	98,250	1,35,270	2.4

T_3	56.99	2,27,968	1,27,000	1,00,968	1.8
T_4	61.00	2,43,984	1,27,000	1;16,984	1.9
T_5	53.30	2,13,200	1,01,780	1,11,420	2.1
T_6	59.13	2,36,520	99,975	1,36,545	2.4
T_7	64.12	2,56,480	97,460	1,59,020	2.6
T_8	72.00	2,88,000	96,875	1,91,125	3.0
T_9	55.62	2,22,480	98,050	1,24,430	2.3
T_{10}	69.38	2,77,536	96,590	1,80,946	2.9
T_{11}	61.20	2,44,800	96,000	1,48,800	2.6
T_{12}	58.42	2,33,680	95,000	1,38,680	2.5

The data furnished in the Table reveal that the application of fresh cow dung extracts and zinc sulphate in T8 resulted in the highest benefit – cost ratios of 3.0 followed by T10 (2.9), T7 (2.6), T11 (2.6), T12 (2.5), T1(2.4), T2 (2.4), T6 (2.4), T9 (2.3) and T5 (2.1). The lowest benefit – cost ratios were obtained in T3 (1.8) and T4 (1.9).

It is obvious that, when compared to other treatments T8 and T10 were low cost and high returns have contributed to the highest cost ratio. The lowest benefit cost ratio in T3 and T4 might be due to the reason that the rate of *Panchakavia* was higher when compared to the other ingredients in the treatments contributed lowest cost ratio in Grand 9 Banana variety.

4.3.2. Benefit Cost Ratio of the treatments in Banana (Robusta Variety)

The analysis of Benefit Cost Ratio of the treatments in Banana (Robusta Variety)is presented in Table 11.

Table 11. Benefit Cost Ratio of the treatments in Banana (Robusta Variety)

Treatments	Yield (t/ha)	Gross Income	Cost of cultivation	Net income	Benefit cost ratio
T_1	57.35	1,72,044	83,750	88,294	2.1
T_2	53.76	1,61,287	83,750	77,537	1.9
T_3	57.35	1,72,044	1,12,500	59,544	1.5
T_4	49.52	1,48,554	1,12,500	36,054	1.3
T_5	52.42	1,57,248	82,300	74,948	1.9
T_6	54.91	1,64,743	85,400	79,343	1.9
T_7	55.81	1,67,427	83,000	84,427	2.0
T_8	60.44	1,81,327	82,375	98,952	2.2

T_9	58.54	1,75,608	83,550	92,058	2.1
T_{10}	56.44	1,69,322	82,875	86,447	2.0
T_{11}	55.54	1,66,622	81,500	85,122	2.0
T_{12}	43.06	1,29,168	80,500	48,668	1.6

A perusal of the data indicated that the application of fresh cow dung extracts and zinc sulphate in T8 resulted in the highest benefit – cost ratio of 2.2 followed by 2.1 in T1 and T9 (2.1), T7 (2.0), T10 (2.0), T11 (2.0), T2 (1.9), T5 (1.9), T6 (1.9), and T12 (1.6). The lowest benefit – cost ratio were obtained in T4 (1.3) and T3 (1.5).

It is evident that, when compared to other treatments T8, T1 and T9 were low cost and high returns have contributed to the highest cost ratio. The lowest cost ratio in T4 and T3 was due to the reason that the rate of Panchakavia was higher when compared to the other ingredients in the treatments that have contributed lowest cost ratio in Robusta Banana variety

4.3.3. Benefit Cost Ratio of the treatments in Banana (Poovan Variety)

The analysis of Benefit Cost Ratio of the treatments in Banana (Poovan Variety) is presented in Table 12.

Table 12. Benefit Cost Ratio of the treatments in Banana (Poovan Variety)

Treatments	Yield (t/ha)	Gross Income	Cost of cultivation	Net income	Benefit cost ratio
T_1	35.55	1,24,425	80,750	43,675	1.5
T_2	31.54	1,10,381	80,750	29,631	1.4
T_3	34.52	1,20,803	1,09,500	11,303	1.1
T_4	27.56	96,469	1,09,500	-13,031	0.9
T_5	29.84	1,04,436	79,280	25,156	1.3
T_6	32.00	1,11,983	82,475	29,508	1.4
T_7	30.56	1,06,969	79,925	27,044	1.3
T_8	36.00	1,26,000	79,375	46,625	1.6
T_9	39.00	1,36,500	80,550	55,950	1.7
T_{10}	30.94	1,08,281	79,090	29,191	1.4
T_{11}	32.34	1,13,203	78,500	34,703	1.4
T_{12}	24.12	84,420	77,500	6,920	1.1

An analysis of the table reveals that the application of fresh cow dung extracts and Borax in T9 resulted in the highest benefit – cost ratio of 1.7 followed by 1.6 in T8, T1 (1.5), T2 (1.4), T6 (1.4), T10 (1.4), T11 (1.4), T5 (1.3), T7 (1.3), T3 (1.1) and T12 (1.1). The lowest benefit – cost ratio obtained in T4 (0.9).

It is evident that, when compared to other treatments T9 and T8 were low cost and high returns have contributed to the highest cost ratio. The lowest cost ratio in T4 was due to the high cost of *Panchakavia* when compare to the other ingredients in the treatments that have contributed lowest cost ratio in Poovan Banana variety.

4.3.4. Benefit Cost Ratio of the treatments in Banana (Red Banana Variety)

The analysis of Benefit Cost Ratio of the treatments in Banana (Red Banana Variety) is presented in Table 13.

Table 13. Benefit Cost Ratio of the treatments in Banana (Red Banana Variety)

Treatments	Yield (t/ha)	Gross Income	Cost of cultivation	Net income	Benefit cost ratio
T_1	52.21	2,34,945	89,750	1,45,195	2.6
T_2	54.00	2,43,000	89,750	1,53,250	2.7
T_3	52.57	2,36,565	1,18,500	1,18,065	2.0
T_4	55.45	2,49,525	1,18,500	1,31,025	2.1
T_5	50.63	2,27,835	88,280	1,39,555	2.6
T_6	51.16	2,30,220	91,475	1,38,745	2.5
T_7	57.20	2,57,400	88,962	1,68,438	2.9
T_8	73.26	3,29,670	88,375	2,41,295	3.7
T_9	69.30	3,11,850	89,550	2,22,300	3.5
T_{10}	69.64	3,13,380	88,087	2,25,293	3.6
T_{11}	51.30	2,30,850	87,500	1,43,350	2.6
T_{12}	44.35	1,99,575	86,500	1,13,075	2.3

The facts provided in the Table revealed that the application of fresh cow dung extracts and zinc sulphate in T8 resulted in the highest benefit – cost ratios of 3.7 followed by 3.6 in T10, T9 (3.5), T7 (2.9), T2 (2.7), T1 (2.6), T5 (2.6), T11 (2.6), T6 (2.5) and T12 (2.3). The lowest benefit – cost ratio were obtained in T3 (2.0) and T4 (2.1).

It is understood that when compared to other treatments T8 and T10 were low cost and high returns have contributed to the highest cost ratio. The lowest cost ratio in T3 and T4 was due to high rate of *Panchakavia* when compared to the other ingredients in the treatments contributed lowest cost ratio in Red Banana variety.

4.4. ATTRIBUTE MEASURES OF THE TREATMENTS IN BANANA

4.4.1.1 Degree of dependability of treatments in Banana by the Scientists

The degree of dependability of treatments in Banana as perceived by the scientists, extensionists and farmers were analysed and depicted in the Table 14. A thorough perusal of the table revealed that highest mean score for scientists was observed in T10 (2.63) followed by T7 (2.53), T8 (2.43), T9 (2.37), T4 (2.23), T3 (2.20), T6 (2.17), T5 (2.1), T2 (2.1) and T1 (2.03). The lowest mean score was obtained for T11 (1.8) and T12 (1.0). The kendall's coefficient of concordance (Kc) (0.484) was significant at 0.01 level. The magnitude of co-efficient indicated that there exists moderate agreement among Scientists with respect to the degree of dependability of treatments in banana since it is less than 0.5.

Table 14. Degree of dependability of treatments in Banana among Scientists, Farmers and Extensionists

Treatment	Degree of dependability in treatments					
	Scientists (n = 30)		Farmers (n = 30)		Extensionists (n = 30)	
	Avg Mean Score	Rank	Avg Mean Score	Rank	Avg Mean Score	Rank
T_1	2.03	IX	2.4	V	1.97	VIII
T_2	2.1	VIII	2.23	VIII	2.03	VII
T_3	2.2	VI	2.5	III	2.27	VI
T_4	2.23	V	2.47	IV	2.4	IV
T_5	2.1	VIII	2.33	VI	2.67	II
T_6	2.17	VII	2.27	VII	2.6	III
T_7	2.53	II	2.9	I	2.33	V
T_8	2.43	III	2.8	II	2.4	IV
T_9	2.37	IV	2.8	II	1.97	VIII
T_{10}	2.63	I	2.8	II	2.73	I

T_{11}	1.8	X	2.27	VII	1.97	VIII
T_{12}	1	XI	1	IX	1	IX

Kendall's coefficient of concordance (Kc) = 0.484 ** Kendall's coefficient of concordance (Kc) = 0.473 ** Kendall's coefficient of concordance (Kc) = 0.485 **
** Significant at 0.01 level; **Significant at 0.01 level; **Significant at 0.01 level

The practice of applying urea, sulphate of potash, water and fresh cow dung in T10 has already been recommended by the Indian Institute of Horticultural Research, Bangalore and hence the scientists of National Agricultural Research System (NARS) would have found this treatment more dependable.

Application of fresh cow dung extract, zinc sulphate and borax which are the main constituents of treatments namely, T7, T8, and T9 have increasingly been debated among the scientists and most of the scientists in the National Research Centre for Banana, (NRCB) Tiruchirapalli have been using these combinations in their experiments.

The lowest scores were obtained in the control (T12) and the application of Cow dung extract solution alone in T11. It is quite natural that the treatments will give good results than the control and in this same line cow dung solution alone may not have yielded good result in the field and hence in the lowest score in the dependability of the treatments in Banana.

4.4.1.2. Degree of dependability of treatments in Banana by Farmers

According to farmers the treatment T7 (2.90) showed the highest mean score followed by T10 (2.80), T9 (2.80) and T8 (2.80), T3 (2.50), T4 (2.47), T1 (2.40), T5 (2.33), T6 and T11 (2.27), T2 (2.23) and the lowest score was observed in T12 (1.00). The kendall's coefficient of concordance (Kc) was (0.473) which is significant at 0.01 level.

The scientists and farmers perception towards the dependability of the treatments was almost similar to each other. The maximum of score was obtained in T7 which is the application of fresh cow dung extract along with zinc sulphate and borax. The result is evident in some banana belt of the study area.

Application of urea, sulphate of potash, water and fresh cow dung in T10 had already found a place in ICAR recommendations. Farmers who were part of the PTD process in the present study had established good contacts with scientists and extension personnel concerned. The combination in T10 and T7 has been a point of discussion among banana growers and some of them have been experimented it in smaller scale. This might have been the reason for their higher dependability score in these treatments.

The kendall's coefficient of concordance (Kc) value for the dependability of farmers (0.473) indicates that there existed moderate agreement among farmers with respect to the different treatments under study. This shows that farmers have similar mind sets as far as the different treatments are concerned.

4.4.1.3. Degree of dependability of treatments in Banana by Extensionists

The highest mean score was recorded by extensionists in T10 (2.73) followed by T5 (2.67), T6 (2.60), T8 (2.40) and T4 (2.40), T7 (2.33), T3 (2.27), T2 (2.03) and the lowest score was obtained in T1 (1.97), T9 (1.97), T11 (1.97) and T12 (1.0). The kendall's coefficient of concordance (Kc) (0.485) was significant at 0.01 level. This implies that there exists a moderate agreement among the extension personnel as far as the dependability scores are concerned.

Most of the extensionists opined that they were extremely confident about the TNAU and ICAR recommendations resulted in the highest score in the dependability of the treatments.

The highest score obtained in the application of urea, sulphate of potash, water and fresh cow dung in T10 and Urea alone in T5 were the ICAR and TNAU recommendations respectively. More over all these recommendations have already been practiced in the farmer's field and also recommended by the extensionists. Thus those treatments get the maximum dependability scores.

The results of the perception of Scientists, farmers and extensionists towards dependability of treatments in banana revealed that the application of urea, sulphate of potash, water and fresh cow dung, and the combinations of fresh cow dung extract, zinc sulphate and borax will be more profitable one in a long run to the banana growers and the field due to the combination of organic and inorganic ingredients.

4.4.2.1. Status of Adoption of treatments in Banana by Scientists

Adoption of agricultural practices/technology is the final stage in adoption process where a farmer becomes convinced about the merits of practice/technology. An account of adoption of components in the selected treatments would be useful to decipher the degree of orientation of stakeholders towards these components. Table 15 gives a vivid picture of mean score of adoption pertaining to the components in different treatments selected for farmers, scientists and extension personnel.

Table 15. Status of Adoption of treatments in Banana among Scientists, Farmers and Extensionists

Treatment	Status of Adoption of treatments in Banana					
	Scientists		Farmers		Extensionists	
	Avg Mean Score	Rank	Avg Mean Score	Rank	Avg Mean Score	Rank
T_1	3.23	VII	3.8	V	2.67	VI
T_2	2.73	XI	3.7	VI	2.37	VIII
T_3	3.07	IX	3.8	V	2.47	VII
T_4	3.4	VI	3.7	VI	3	III
T_5	4.47	I	4	III	4.13	I
T_6	4.3	II	4	III	3.4	II
T_7	3.6	IV	4	III	2.47	VII
T_8	3.47	V	3.8	V	2.47	VII
T_9	3.17	VIII	3.9	IV	2.37	VIII
T_{10}	4.1	III	4.07	I	2.77	V
T_{11}	2.9	X	4.03	II	2.8	IV
T_{12}	1	XII	1.4	VII	1.1	IX

Kendall's coefficient of concordance (Kc) = 0.519 **Kendall's coefficient of concordance (Kc) = 0.493 ** Kendall's coefficient of concordance (Kc) = 0.502 **
** Significant at 0.01 level ** Significant at 0.01 level ** Significant at 0.01 level

The highest mean score as perceived by scientists was T5 (4.47) followed by T6 (4.30), T10 (4.10), T7 (3.60), T8 (3.47), T4 (3.40), T1 (3.23), T9 (3.17), T3 (3.07), T11 (2.90) and T2 (2.73). The lowest mean score was observed in T12 (1.00). The kendall's coefficient of concordance (Kc) (0.519) was significant at 0.01 level. This revealed that the scientists had an appreciable agreement among themselves in ranking the treatments.

The scientists opined that the farmers adopted the TNAU recommendations such as application of urea alone for a long time and hence placed the highest score.

In the scientist's opinion, application of Urea, Sulphate of Potash, water and fresh cow dung and application of fresh cow dung extract, zinc sulphate and borax and their combinations were more prominently adopted by farmers on a sustained basis so the scores were high when compared to others.

4.4.2.2. Status of Adoption of treatments in Banana by Farmers

The mean scores were worked out to find out the status of adoption as perceived by the farmers and are presented in Table 15. T10 (4.07) has gained the highest mean score followed by T11 (4.03), T5 (4.00), T6 (4.00), and T7 (4.00), T9 (3.90), T1 (3.80), T3 (3.80) and T8 (3.80), T2 (3.70) and T4 (3.70). The lowest mean score was obtained in T12 (1.40). The kendall's coefficient of concordance (Kc) (0.493) was significant at 0.01 level, which indicates that a moderate level of agreement was found among the farmers with respect to adoption of practices.

The highest score acquired in T10 (i.e.) application of urea, sulphate of potash, water and fresh cow dung may be due to the fact that a good number of the farmers in the banana belt were practicing the technology as it produced good results, and a recommended one.

Majority of the farmers in the study area said that spraying of cow dung solutions in banana is a common practice and specifically it is used for tying in the bunch to get uniform growth of bunches. Nowadays it is common among banana growers. This might have been the reason for the applications of Cow dung extract solution in T11 obtain to higher scores in the status of adoption of the treatments in banana.

The recommendation of TNAU for the uniform growth of bunches is the tying of Urea which is being practiced widely among the banana growers and it has got a higher score in the status of adoption of the treatments in Banana.

Application of urea and sulphate of potash and fresh cow dung extract, zinc sulphate and borax are recently introduced practices in the field condition paving a way to a higher score in the status of adoption of the treatment in Banana.

4.4.2.3. Status of Adoption of treatments in Banana by Extensionists

The highest mean score as perceived by the extensionists was T5 (4.13) followed by T6 (3.40), T4 (3.00), T11 (2.80), T10 (2.77), T1 (2.67), T3 (2.47), T7 (2.47) and T8 (2.47), T2 (2.37) and T9 (2.37). The lowest score was recorded in T12 (1.10). The kendall's coefficient of concordance (Kc) (0.502) was significant at 0.01 level, which reveals that a high degree of agreement exists between extension personnel in rating the treatments for their adoption in the field.

The most promising practice among the banana farmers is the tying of Urea in the bunch of banana. The banana farmers have been adopting the technology for a long period as it was recommended by the extensionists.

Application of urea and Sulphate of Potash combinations has many advantages which prompted the extensionists to advocate the combinations to the banana farmers as a successful one. Moreover these combinations of the treatments are already been practised by the farmers in the study area. This resulted in high score with regard to the status of adoption of the treatments in banana.

Of late the extensionists are advocating organic – driven technologies such as, *Panchakavia*. The combinations like *Panchakavia* and household wood ash were gaining momentum in the farmer's field as well as from the extensionist's perspectives and these combinations were tried in the farmer's fields. Hence the score was high in this aspect of status of adoption of treatments in banana.

To wrap up, the scientists, farmers and extensionists point of view towards the status of adoption of treatments in banana emphasized that the TNAU and IIHR recommendations viz., application of urea alone and application of urea, sulphate of potash, water and fresh cow dung respectively played a pivotal role in the adoption of the aforesaid combinations by the farmers.

4.4.3.1. Assignment of weightage score to treatments in Banana by Scientists

The weightage scores assigned to the treatments by scientists, farmers and extensionists were explored and are presented in Table 16. The highest mean score of scientists was obtained in T10 (8.70) followed by T8 (7.47), T7 (7.37), T9 (7.17), T6 (6.83), T5 (5.73), T4 (5.3), T3 and T11 (5.27), T1 (5.1) and T2 (4.73). The lowest mean score was observed in T12 (0.33). The Kendall's coefficient of concordance (Kc) (0.586) was significant at 0.01 level.

Table 16. Assignment of Weightage score to treatments in Banana among Scientists, Farmers and Extensionists

Treatment	Assignment of Weightage score to treatments					
	Scientists		Farmers		Extensionists	
	Avg Mean Score	Rank	Avg Mean Score	Rank	Avg Mean Score	Rank
T_1	5.1	IX	6.33	VIII	5.9	X
T_2	4.73	X	6	X	5.47	XI
T_3	5.27	VIII	6.23	IX	6.4	VIII

T_4	5.3	VII	6.4	VII	7.1	VI
T_5	5.73	VI	6.53	VI	7.83	II
T_6	6.83	V	7.17	V	7.43	III
T_7	7.37	III	8.27	III	7.13	V
T_8	7.47	II	8.2	IV	7.2	IV
T_9	7.17	IV	8.33	II	6.2	IX
T_{10}	8.7	I	8.6	I	8.17	I
T_{11}	5.27	VIII	5.87	XI	6.57	VII
T_{12}	0.33	XI	0.23	XII	0.77	XII

Kendall's coefficient of concordance (Kc) = 0.586 **Kendall's coefficient of concordance (Kc) = 0.559 ** Kendall's coefficient of concordance (Kc) = 0.464 **
** Significant at 0.01 level ** Significant at 0.01 level **Significant at 0.01 level

The recommendation of IIHR viz., the application of urea, sulphate of potash, water and fresh cow dung is created more impact among banana growers and the scientists. Moreover, it was a proven technology and has obviously been assigned highest score by the scientists.

Of late, the scientists working in banana are convinced with the combinations of treatments like application of fresh cow dung extract, zinc sulphate and borax. Which is reflected in assigning higher scores to these treatment combinations.

It could be observed from the table that Kc value (0.586) for all the treatments and the assignment of weightage score to the treatments by the scientists were significant at one per cent level of probability indicates that there is no significant difference between the treatments and the assignment of weightage score to the treatments as perceived by the scientists.

From the overall view point of the scientists, the highest scores assigned to the treatment combinations like application of urea, sulphate of potash, water and fresh cow dung and fresh cow dung extract, zinc sulphate and borax and their combinations will be highly beneficial to the banana farmers.

4.4.3.2. Assignment of weightage score to treatments in Banana by Farmers

The farmers weightage score to the treatments have been exhibited in table and revealed that highest mean score was attained in T10 (8.6) followed by T9 (8.33), T7 (8.27), T8 (8.2), T6 (7.17), T5 (6.53), T4 (6.4), T1 (6.33), T3 (6.23), T2 (6.0) and T11 (5.87). The lowest mean score was recorded in T12 (0.23). The Kendall's coefficient of concordance (Kc) (0.559) was significant at 0.01 level.

The farmer's assignment of weightage score to the treatments in banana is almost similar to that of the scientists. This showed the effectiveness and utility of the treatments in all spheres.

The farmers in the study area assigned the highest score to the application of urea, sulphate of potash, water and fresh cow dung which is an IIHR recommendation.

The farmers are also convinced about the usefulness of the application of fresh cow dung extract, zinc sulphate and borax and their combinations similar to the scientist perspectives. These treatment combinations perform well in the field conditions and had been reported by all the three systems viz., RSS, FSS and ESS. So evidently the farmers assigned the high scores to these combinations

It is evident from the table that Kc value (0.559) for all the treatments and the assignment of weightage score to the treatments by the farmers were significant at 1 per cent level of probability. This indicates that there is no significant difference between the treatments and the assignment of weightage score to the treatments as perceived by the farmers.

With regard to the assignment of weightatge score by farmers showed that the treatments like application of urea, sulphate of potash, water and fresh cow dung and the application of fresh cow dung extract, zinc sulphate and borax and their combinations is gaining momentum in recent years and definitely it will get more impact in the field conditions. The treatment combinations expressed by the farmers have been already practiced by some of the farmers in their field.

4.4.3.3. Assignment of weightage score to treatments in Banana by Extensionists

The highest mean score to the treatments assigned by the extensionists was in T10 (8.170 followed by T5 (7.83), T6 (7.43), T8 (7.2), T7 (7.13), T4 (7.1), T11 (6.57), T3 (6.4), T9 (6.2), T1 (5.9) and T2 (5.47). The lowest mean score was obtained in T12 (0.77). The Kendall's coefficient of concordance (Kc) (0.464) was significant at 0.01 level.

Application of urea, sulphate of potash, water and fresh cow dung combination was assigned with highest score by the extensionists and found to be more promising technology in the field which is an IIHR recommended practice. Its effectiveness was experienced by the extensionists in the field conditions as a cost effective one.

The tying of urea alone was assigned with high score by the extensionists due to the fact that the technology is adopted by the most of the farmers in the study area. The farmers were expressing the worth of the treatment in a positive manner, which may lead the extensionists to assign high score

The extensionists are recommending the combination of urea and sulphate of potash so as to get a uniform growth of bunches in banana. These combinations performed well in the field conditions as reported by the farmers to the extensionists that paved a way to assign high score the aforesaid treatment in banana.

A glimpse at the table revealed that the Kc value (0.464) for all the treatments and the assignment of weightage score to the treatments by the extensionists were significant at 1 per cent level of probability. It indicates that there is no significant difference between the treatments and the assignment of weightage score to the treatments by the extensionists

The extensionists point of view in relation to the assignment of score to the treatments exposed that the application of urea, sulphate of potash, water and fresh cow dung, tying of urea alone as such and application of urea and sulphate of potash were the most opt and fitted technology in the field conditions and the money spent to these tcchnologies were minimal. These technologies would become popular in near future to get more yield and income.

The overall views of scientists, farmers and extensionists in assigning scores to the treatment combinations in banana were almost similar to each other. They assigned the highest scores to the IIHR recommendation and the combinations of fresh cow dung extract, zinc sulphate and borax application. These technologies perform well in the field conditions and are economically feasible.

4.4.4.1. Evaluative perception of treatments in banana by Scientists

The Evaluative perception of treatments in banana as perceived by the scientists, extensionists and farmers were gathered, analyzed and depicted in the Table 17. A perusal of the table revealed that highest mean score for scientists was observed in T10 (4.63), T7 (4.33), T8 (4.3), T9 (4.23), T6 (4.2), T4 (3.77), T5 (3.73), T1 and T3 (3.7), T2 (3.6) and T11 (3.37). The lowest mean score was attained in T12 (1.2).

The Kendall's coefficient of concordance (Kc) (0.491) was significant at 0.01 level.

Table 17. Evaluative Perception of treatments in Banana among Scientists, Farmers and Extensionists

Treatment	Evaluative Perception of treatments in Banana					
	Scientists		Farmers		Extensionists	
	Avg Mean Score	Rank	Avg Mean Score	Rank	Avg Mean Score	Rank
T_1	3.7	VIII	4.1	VI	3.97	VI
T_2	3.6	IX	4	VII	3.93	VII
T_3	3.7	VIII	4.1	VI	4.17	IV
T_4	3.77	VI	4.13	V	4.23	III
T_5	3.73	VII	4.13	V	4.23	III
T_6	4.2	V	4.3	IV	4.23	III
T_7	4.33	II	4.7	III	4.17	IV
T_8	4.3	III	4.77	II	4.4	II
T_9	4.23	IV	4.77	II	4.07	V
T_{10}	4.63	I	4.8	I	4.77	I
T_{11}	3.37	X	3.93	VIII	3.97	VI
T_{12}	1.2	XI	1	IX	1.2	VIII

Kendall's coefficient of concordance (Kc) = 0.491 **Kendall's coefficient of concordance (Kc) = 0.499 ** Kendall's coefficient of concordance (Kc) = 0.496 **
** Significant at 0.01 level ** Significant at 0.01 level ** Significant at 0.01 level

The scientist's evaluative perception of treatments in banana indicated that the application of urea, sulphate of potash, water and fresh cow dung in (T10) rated as highest one due to the fact that the technology is pretty powerful, proven and recommended by the IIHR. According to scientists perspectives that the aforesaid combination was inevitable to get more yield and uniform growth of bunches from top to bottom and very cost effective.

The application of fresh cow dung extract, zinc sulphate and borax and their combinations were the most field oriented combinations in all the aspects and moreover the results were quite convincing. The farmers in the study area applied the technology in their field so as to get more yield and income. Thus the scientists have evaluated the technology with high score.

It could be observed from the table that Kc value (0.491) for all the treatments and the evaluative perception of the treatments by the scientists were significant at 1% level of probability. It indicates that there is no significant difference between the treatments and the evaluative perception of the treatments as perceived by the scientists.

4.4.4.2. Evaluative perception of treatments in banana by Farmers

The farmers evaluative perception of treatments in banana were analysed and it revealed that the highest mean score as per farmers point of view was observed in T10 (4.8) followed by T8 and T9 (4.77), T7 (4.7), T6 (4.3), T4 and T5 (4.13), T1 and T3 (4.1), T2 (4.0) and T11 (3.93). The lowest mean score was attained in T12 (1.0). The Kendall's coefficient of concordance (Kc) (0.499) was significant at 0.01 level.

Application of urea, sulphate of potash, water and fresh cow dung in T10 has been evaluated as the highest one with respect to the farmers perceptive. The reason behind that it is an established technology and the farmers have been satisfied with the result of more yield, uniform growth of bunches from top to bottom, keeping quality also good more over it's a cost effective technology, so it is popular among the farming community.

The farmers in the study area were well pleased with the treatment combinations of fresh cow dung extract, zinc sulphate and borax. These combinations were tried by the farmers in the study area and the results of the treatment combinations were quite realistic. These facts would have motivated the farmers to evaluate high to the treatment combinations.

It is evident from the table that Kc value (0.499) for all the treatments and the evaluative perception of the treatments by the farmers were significant at 1 per cent level of probability. It indicates that there is no significant difference between the treatments and the evaluative perception of the treatments as perceived by the farmers.

The farmers evaluative perception of the treatments in banana totally rely on the application of urea, sulphate of potash, water and fresh cow dung and the application fresh cow dung extract, zinc sulphate and borax and their combinations resulted as a realistic one. Moreover all the technologies were cost effective and organic which is a gifted one to the future generation and the forth coming days of the banana growers.

4.4.4.3. Evaluative perception of treatments in banana by Extensionists

Extensionists evaluative perception of treatments in banana were evaluated and the highest mean score was observed in T10 (4.77) followed by T8 (4.4), T4, T5 and T6 (4.23), T3 and T7 (4.17), T9 (4.07), T1 and T11 (3.97), T2 (3.93). The lowest mean score was obtained in

T12 (1.2). The Kendall's coefficient of concordance (Kc) (0.496) was significant at 0.01 level.

Extensionists evaluated the treatments as similar to that of scientists and farmers as application of urea, sulphate of potash, water and fresh cow dung due to the reason that it's a IIHR recommendation, proven one and cost effective. The scientists and extensionists are recommending the technology to the farmers and attributing the highest score for the technology.

Application of fresh cow dung extract and zinc sulphate is one of the most emerging technologies adopted by the farmers and it has been identified as an effective one by the extensionists to advocate to the banana growers as a successive one.

Application of *panchakavia* and household wood ash is the combination tried by most of the farmers in the study area due to the reason that it's an organic one and cost effective. Now-a-days extensionists are recommending the *panchakavia* to the farmers and hence evaluated the combinations with the highest score.

The TNAU recommendation of tying of urea has also been evaluated by the extensionists with high score because as it has been practiced in the field conditions for a long time, so the extensionists rated the treatment as the highest one.

Application of Urea and sulphate of Potash is a positive combination in the field conditions and used by the farmers of the study area. Though it is inorganic it has produced more yield hence it has been extensively used by the farmers and paved a way to the extensionists to score high.

The table further revealed that the Kc value (0.496) for all the treatments as perceived by the extensionists were significant at 1% level of probability indicated that no significant difference exists between the treatments.

The extensionist's perception towards the treatments in banana reveals that application of urea, sulphate of potash, water and fresh cow dung, fresh cow dung extract and zinc sulphate, Panchakavia and household wood ash, tying of urea and Urea and sulphate of Potash were appropriate to the field conditions and give more yield with low expenses. It could be easily accepted and adopted by any category of the farming community.

To sum up the evaluative perception of scientists, farmers and extensionists towards treatments in banana and their assignment of scores to the treatments are on par with each other. They evaluated the application of urea, sulphate of potash, water and fresh cow dung and the combinations of fresh cow dung extract, zinc sulphate and borax were most appropriate to the field conditions, resulting in more yield and the expenses were very low. It could be very well be adopted by the farming community for good results.

4.4.5. 1. Rationalization of treatments in Banana by Scientists

The Rationalization of treatments in Banana as perceived by the scientists and extensionists were collected, analysed and represented in Table 18. A cursory look at the table revealed that highest means score for scientists was observed in T10 (4.47) followed by T7 (4.3), T8 (4.23), T6 (4.2), T9 (4.1), T5 (3.93), T3 and T4 (3.87), T1 (3.83), T2 (3.7) and T11 (3.57). The lowest mean score was attained in T12 (1.0). The Kendall's coefficient of concordance (Kc) (0.527) was significant at 0.01 level.

It is obvious that the application of urea, sulphate of potash, water and fresh cow dung was highly rationalized by the scientists due to the reason that it is an IIHR recommendation, gives more yields, cost effective with uppermost scientific background and a proven technology. This base lifts up the technology as a rationalized one among the scientists.

Table 18. Rationalization of treatments in Banana among Scientists, Farmers and Extensionists

Treatment	Rationalization of treatments in Banana			
	Scientists		Extensionists	
	Avg Mean Score	Rank	Avg Mean Score	Rank
T_1	3.83	VIII	4.03	VI
T_2	3.7	IX	4	VII
T_3	3.87	VII	4.33	IV
T_4	3.87	VII	4.43	III
T_5	3.93	VI	4.57	I
T_6	4.2	IV	4.43	III
T_7	4.3	II	4.1	V
T_8	4.23	III	4.33	IV
T_9	4.1	V	4	VII
T_{10}	4.47	I	4.5	II
T_{11}	3.57	X	4.1	V
T_{12}	1	XI	1.07	VIII

Kendall's coefficient of concordance (Kc) = 0.527; **Kendall's coefficient of concordance (Kc) = 0.527**; ** Significant at 0.01 level ** Significant at 0.01 level

The other treatment combinations were rationalized in the order of importance by the scientists due to the reason that they were coming up well in the field conditions with respect to more yield, uniform growth of bunches, more scientific reason, cost effective and can be followed by all the categories of the farmers.

It is evident from the table that Kc value (0.527) for all the treatments and the rationalization of the treatments by the scientists were significant at 1 per cent level of probability. It indicates that there is no significant difference between the treatments and the rationalization of the treatments as perceived by the scientists.

4.4.5.2. Rationalization of treatments in Banana by Extensionists

The extensionist's rationalization of treatments in banana were gathered. The highest mean score was obtained in T5 (4.57) followed by T10 (4.5), T4 and T6 (4.43), T3 and T8 (4.33), T7 and T11 (4.1), T1 (4.03), T2 and T9 (4.0). The lowest mean score was attained in T12 (1.07). The Kendall's coefficient of concordance (Kc) (0.527) was significant at 0.01 level.

The extensionist's rationalization was somewhat similar to that of the scientists. The extensionists rationalized the TNAU and IIHR recommendations as tying of urea alone and application of urea, sulphate of potash, water and fresh cow dung respectively. These treatments were highly rationalized because of its accessibility, suitability, scientific background, more yield and cost effective. All other treatments except control were rationalized by the extensionists due to the said reason.

A glimpse at the table revealed that the Kc value (0.527) for all the treatments and the rationalization of the treatments by the extensionists were significant at 1 per cent level of probability indicated that there is no significant difference between the treatments and the rationalization of the treatments as perceived by the extensionists

To sum up the rationality of that scientists and extensionists were almost similar to each other keeping the view that all the treatments except control were highly suited to the field conditions, more scientific background, more yield and cost effective. It is evident from the data that the treatments taken for the study is highly useful, indispensable, successive one to the banana growers to increase their farm income and to ultimately improve the standard of living.

4.5. COMPARATIVE ANALYSIS OF ATTRIBUTE MEASURES OF THE TREATMENTS IN BANANA AMONG RESEARCH SUPPORT SYSTEM (RSS), FARMERS SUPPORT SYSTEM (FSS) AND EXTENSION SUPPORT SYSTEM (ESS)

The comparative analysis on attribute measures of the treatments in banana among Research Support System, Farmers Support System and Extension Support System was worked out using Spearman's rank order correlation coefficient (rs) to know the degree of relationship among the three systems. The results are summarized in Table 19.

4.5.1. Comparative analysis of Degree of dependability of treatments in Banana

The findings of spearman rank order correlation from the table revealed that there was no significant relationship between FSS and ESS with respect to Degree of dependability in the treatments in Banana. The farmers and extensionists point of view regarding the dependability of treatments in banana was not related to each other. Farmers were more concerned about the treatment combinations which give high yield, minimum expenses and easy availability of raw materials. Whereas extensionists were concerned more with the recommendations forwarded by an authentic credible agency.

Table 19. Spearman's rank correlation co efficient of attribute measures of the treatments in banana among Research Support System, Farmers Support System and Extension Support System.

System	Degree of dependability in treatments	Status of adoption of treatments	Assignment of weightage score to treatments	Evaluative Perception of Treatments	Rationalization of Treatments
FSS and ESS	0.349 NS	0.536 NS	0.664*	0.792 **	Nil
FSS and RSS	0.887**	0.586 *	0.932 **	0.792 **	Nil
RSS and ESS	0.594*	0.709 **	0.802 **	0.789 **	0.587 *

* Correlation is significant at the.05 level (2-tailed); ** Correlation is significant at the 0.01 level (2-tailed); NS – Non Significant; FSS – Farmers Support System; RSS – Research Support System; ESS – Extension Support System

Positive and significant relationship exists between FSS and RSS with respect to the degree of dependability in treatments in Banana at one per cent level. The dependability of treatments in banana as perceived by the scientists and farmers were almost alike. The farmers and scientists depended on the treatments which is practically feasible, economically viable and eco friendly.

There exists a positive and significant relationship between RSS and ESS at five per cent level with respect to the degree of dependability in treatments. The dependability of treatments in banana as perceived by the scientists and extensionists were more or less similar to each other which are explained elsewhere. They totally rely on the recommendations by the agency and also placed importance to the treatment combinations of higher yield, easily availability of raw materials and practice, less/no cost and so on. The scientists and extensionists nowadays extremely depend on the organic based treatments due to the farmer's preferences to the organic, which was projected in the dependability scores for treatments in banana.

4.5.2. Comparative analysis of Status of adoption of treatments in Banana

The findings of spearman rank correlation reveal that there was no significant relationship between FSS and ESS pertaining to status of adoption of treatments in Banana. The status of adoption of treatment combinations of farmers and extensionists were not similar to each other. When the farmers adopted the technology, it must give more yield, less or no cost, locally available raw materials and easy to practice. The technology which suits the aforesaid conditions would naturally be adopted by the farmers. At the same time the extensionist's approach will be focused on to the already practised one and the recommendations of the agency reflected in the status of adoption of the treatment combinations in banana.

Positive and significant relationship between FSS and RSS was observed at five per cent level pertaining to status of adoption of treatments in Banana. The status of adoption of treatments in Banana by the farmers and the scientists were somewhat similar to each other. They were very clear in adopting the technologies recommended by the agencies. At the same time they have given due importance to the technologies which are organic. Most of the farmers used cow dung extract irrespective of the crops they grow. The importance of Panchakavia, cow based ingredients were equally perceived by the scientist and farmers resulted in adoption of the treatment combinations in Banana by both groups.

The data generated in the table exhibited that there was positive and significant relationship between RSS and ESS at one per cent level pertaining to the status of adoption of treatments. With regard to the status of adoption of treatments in banana, the opinion of scientists and extensionists were alike. They have recommended the technologies which will be highly useful to the farming community, was reflected in the status of adoption. The scientists and extensionists were clear in their job to recommend the technologies of the agencies due to the reason that it is a proven technology. Scientists and extensionists were aware of the inevitability of recommending the organic technology to the banana growers, in recent times.

4.5.3. Comparative Analysis of Assignment of weightage score to treatments in Banana

The findings of spearman rank correlation revealed a positive and significant relationship between FSS and ESS only at five per cent level as far as the assignment of weightage score to treatments in Banana. The farmers and extensionists were slightly differing from each other in the assignment of weightage score to treatments in banana due to the fact that the performance of the particular technology within the perspective of farmers and the extensionists will differ from each other.

Farmers and Extensionists assigned highest score to the IIHR recommendations because of their superior performance of the practices in the field conditions. Farmers were satisfied with the results of the fresh cow dung extract, zinc sulphate and borax combinations in the field conditions but at the same time the extensionists assigned high score to the TNAU recommendations as they are recommending the same. When compared to the extensionists the farmers were pleased with Panchakavia, wood ash and cow urine which are locally available and being used by them. But the extensionists were not convinced as it has not scientifically proved. Also their exposure to the treatment combinations was very minimal.

There was positive and significant relationship between FSS and RSS at 1% level in the assignment of weightage Score to treatments in Banana. The results of the treatment combinations in banana experienced by the scientists and farmers were alike and thus the assignment of weightage score to the treatment combinations is almost similar. The farmers and scientists were happy with the results of IIHR recommendations. At the same time the fresh cow dung extract, zinc sulphate and borax combinations, Panchakavia, Ash and cow urine also resulted in more yield and quiet impressive with both the category. The results were almost similar due to the practicability, economic viability and eco friendliness of the technology.

The data given in table exhibited that there was positive and significant relationship between RSS and ESS at one per cent level as far as assignment of weightage Score to treatments in Banana are concerned. The scientists and extensionists were more specific in assigning the score to the treatment combinations and similar to each other. They are comfortable in their job to assign high score to IIHR recommendations and TNAU. Scientists and Extensionists are recommending the technology which is well suited to the field conditions and more benefited to the farmers in all the spheres. This was reflected in both of them similarly in assigning highest score.

4.5.4. Comparative Analysis of Evaluative Perception of Treatments in Banana

The findings of spearman rank correlation revealed that there was positive and significant relationship between FSS and ESS at one per cent level with respect to evaluative perception of the treatments in banana. The Extensionist's and farmer's evaluative perception towards the treatments in banana was on par with each other that means there in no deviation in evaluating the treatments. Farmers and extensionists were satisfied with the recommendations of IIHR, Bangalore and the results were apparent. The prominent and recent one viz., fresh cow dung extract, zinc sulphate and borax combinations was practically proven one in the field conditions which might have motivated the farmers and extensionists to evaluate next to the established recommendations. Moreover, they were positive towards TNAU recommendations, Panchakavia, wood ash and cow urine.

There was positive and significant relationship between FSS and RSS at one per cent level with respect to evaluative perception of the treatments in banana. Farmers and scientist were almost parallel in evaluating the treatment combinations in banana. They evaluated the recommendations of the IIHR as high because of the proven nature. Besides they evaluated the other treatment combinations similar due to the fact that they are practically feasible, economically viable, eco and farmer friendly. There was no deviation among the treatment combinations between the farmers and scientists due to the reason that the observability of the technology was experienced by both the category of the people.

The data recorded in table exhibited that there was positive and significant relationship between RSS and ESS at one per cent level. The scientists and extensionists perceived and evaluated the treatments combination in banana almost in a similar way. The scientists and extensionists are more concerned with the proven technology and its suitability in the field and hence evaluated the recommendations of IIHR as a higher one.

The other treatment combinations were evaluated by the scientists and extensionists based on the faith, performance and practical feasibility of the technology. They were convinced about the technologies like fresh cow dung extract, zinc sulphate and borax combinations, Panchakavia and ash. In the extensionist's point of view, the TNAU recommendations are practiced more in the field conditions, but scientists differ in their point of view that they are not satisfied with the application of urea alone in banana and at present they are not recommending the technology to the farmers.

4.5.5. Comparative analysis of scientific rationality of treatments in Banana

The findings of spearman rank correlation revealed that there was positive and significant relationship between RSS and ESS only at five per cent level pertaining to rationality of treatments in banana. The scientists and extensionists were very clear in rationalizing the treatments in banana. They were highly confident about the recommendations given by IIHR and TNAU due to the fact that they are proven technologies and no need to probe further. With regard to the other treatment combinations they differ in their rationalization based upon their experiences.

The scientists rationalized the combinations of fresh cow dung extract, zinc sulphate and borax than the organic one namely *panchakavia,* ash and cow urine combination due to the reason that they were much satisfied with the performance of fresh cow dung extract, zinc sulphate and borax combinations in the field. *Panchakavia* is now getting popular in the field and the scientific rationale behind the *panchakavia* and cow urine combinations are yet to be fully ascertained by the scientists.

The extensionists rationalized the treatment combinations of both organic and inorganic and give due importance to the performance of the technology in the field conditions and the adoption of the technology by the banana growers. Extensionists were very much confident about the panchakavia and its uses. The treatment combinations like fresh cow dung extract, zinc sulphate and borax were test verified by the extensionists in the field conditions and its results are evident which made the extensionists to rationalize the treatments.

4.6. PERCEPTION OF SCIENTISTS, EXTENSIONISTS AND FARMERS TOWARDS PARTICIPATORY TECHNOLOGY DEVELOPMENT (PTD)

4.6.1. Perception of Scientists towards Participatory Technology Development (PTD)

Perceived opinion of any psychological object would pave way for positive/negative attitude towards the objects. A study of perceived opinion of PTD by all stakeholders of PTD namely, Farmers/ Scientists/Extension agent would unravel the degree of agreement towards different dimension of PTD. The data collected pertaining to perceived opinion of PTD by Scientists/Farmers/Extension personnel is presented.

Table 20. Perception of Scientists towards to Participatory Technology Development (PTD) (n=30)

Sl. No	Statement	Perception toward PTD			
		Agree		Disagree	
		No	%	No	%
1.	PTD is a collaborative activity involving farmers, scientists and extensionists for identification, appreciation, rectification of farmers innovative ideas	28	93.33	2	6.67
2.	PTD is the incorporation of farmers knowledge into technology design and evaluation	29	96.67	1	3.33
3.	Knowledge sharing between farmers and scientists/extensionists	28	93.33	2	6.67
4.	PTD is a multi disciplinary approach and communication between farmer to farmer, agency to farmers and scientists to farmers	29	96.67	1	3.33
5.	PTD is a farmers participatory research with interaction between scientists/extensionists and farmers to test the technology.	25	83.33	5	16.67

The data on Perception of scientists towards the Participatory Technology Development (PTD) were collected and depicted in Table 20.

From the table, it could be inferred that all the perception statements were agreed by the scientists except a few who disagreed the perception statements. More than 90.00 per cent of the scientists agreed with the statements viz., "PTD is the incorporation of farmer's knowledge into technology design and evaluation", "PTD is a multi disciplinary approach and communication between farmer to farmer, agency to farmers and scientists to farmers", "Knowledge sharing between farmers and scientists/extensionists" and "PTD is a collaborative activity involving farmers, scientists and extensionists for identification, appreciation, rectification of farmers innovative ideas" followed by the 83.33% with "PTD is a farmers participatory research with interaction between scientists/ extensionists and farmers to test the technology".

Majority (93.33%) of scientists agreed the statement "PTD is a collaborative activity involving farmers, scientists and extensionists for identification, appreciation, rectification of farmers innovative ideas" and "Knowledge sharing between farmers and scientists/ extensionists". This opinion paves that the scientists did realize the 'sharing of information' is the pre requisite for PTD process. This finding emboldens the perceived opinion of scientists that farmers are co-researchers whose innovations are to be given due recognition. Majority (96.67%) of scientists agreed the statements namely "PTD is the incorporation of farmers knowledge into technology design and evaluation" and "PTD is a multi disciplinary approach and communication between farmer to farmer, agency to farmers and scientists to farmers". The crux of these statements revealed that scientists did favour the linkage between farmers themselves and scientists, which see the process of PTD cycle completes perfectly without any misconceptions. The idea of incorporation of farmer's innovations which they inherited from their forefathers or synthesized by them through smaller experimentation in their field was high in the minds of scientists. From their opinion it is seen that the scientists are fully for the idea of including the farmer's innovations in their research protocol.

Majority (83.33) Scientists agreed the perception statement "PTD is a farmers participatory research with interaction between scientists/extensionists and farmers to test the technology". It is to be noted that unlike other perception statements, this statement evoked 'disagree' opinion among nearly one fifth (16.67%) of scientists who were interviewed. The scientists who disagreed had a notion that it is difficult to establish a participatory research platform in near future.

4.6.2. Perception of Extension Personnel towards Participatory Technology Development (PTD)

Extension personnel are the vital link between Scientists and farmers. Synonymously they are change agents who bring about change in the cognitive and affective domains of clientele i.e. farming community. Their facilitation in taking the process of PTD to the farmers is very much needed since they are more proximate to the farmers than scientists. Their perception towards PTD would definitely influence the diffusion of the concept among farmers. Since, their perception towards PTD was also studied and the results are presented in Table 21.

The results from the table shows that cent per cent of extension personnel who were interviewed agreed all the perception statements. This shows the belief and commitment of Extension personnel towards PTD. Among the perception statements, "PTD is a collaborative activity involving farmers, scientists and extensionists for identification, appreciation, rectification of farmers innovative ideas", "PTD is the incorporation of farmers knowledge into technology design and evaluation", "Knowledge sharing between farmers and scientists/ extensionists", "PTD is a multi disciplinary approach and communication between farmer to farmer, agency to farmers and scientists to farmers", and "PTD is a farmers participatory research with interaction between scientists/extensionists and farmers to test the technology" were endorsed by the Extension personnel. These results exhibit the extension personnel's clear understanding about PTD that sharing of information, incorporation of farmer's knowledge in research design, a multidisciplinary perspective among three systems namely Scientists, farmers and Extension Personnel.

Table 21. Perception of Extension Personnel towards Participatory Technology Development (PTD) (n=30)

S.N	Statement	Perception toward PTD	
		Agree	
		No	%
1.	PTD is a collaborative activity involving farmers, scientists and extensionists for identification, appreciation, rectification of farmers innovative ideas	30	100.0

2.	PTD is the incorporation of farmers knowledge into technology design and evaluation	30	100.0
3.	Knowledge sharing between farmers and scientists/extensionists	30	100.0
4.	PTD is a multi disciplinary approach and communication between farmer to farmer, agency to farmers and scientists to farmers	30	100.0
5.	PTD is a farmers participatory research with interaction between scientists/extensionists and farmers to test the technology.	30	100.0

Extension personnel's perception towards these statements were indicative of their caveats about the 'collaboration' and 'participation' of farmers in research despite being convinced with the potential of PTD.

4.6.3. Perception of Farmers towards Participatory Technology Development (PTD)

Farmers, being an indispensable stakeholder in the process of PTD, their perceived opinion towards PTD must influence them to be a part of this process. Hence their perceived opinion towards PTD was assessed and presented in Table 22.

A thorough look into Table 22 revealed that cent per cent of respondents were fully aware of the benefits of PTD as a process of bringing agricultural science and scientist closer to them. All the perception statements were "fully agreed" by the sample farmers which indicated their favourable attitude towards PTD.

Table 22. Perception of Farmers towards Participatory Technology Development (PTD) (n=30)

Sl. No	Statement	Perception toward PTD			
		Agree		Disagree	
		No	%	No	%
1.	PTD is a farmers scientists collaboration in research	29	96.67	1	3.33
2.	PTD is a farmer to farmer network	30	100.0	0	0
3.	PTD is strengthening farmers innovative capacities	30	100.0	0	0
4.	PTD is satisfactory information exchange between farmers and researchers	30	100.0	0	0
5.	PTD is a knowledge contribution by scientists/extensionists and farmers.	30	100.0	0	0

Cent per cent of respondents agreed with the statements namely "PTD is a farmer to farmer network", "PTD is strengthening farmers innovative capacities", "PTD is satisfactory information exchange between farmers and researchers" and "PTD is a knowledge contribution by scientists/extensionists and farmers. These statements reaffirm the potential of PTD as a facilitator for satisfactory information exchange which in turn results in knowledge contribution in a two – way channel. Further the statements conveyed the farmer's strong belief over farmer to farmer net work. It is obvious that any individual will get convinced, if other member of peer group does something for a good cause. PTD is a solid example of how the concept gets diffused among social system.

The statement "PTD is a farmers scientists collaboration in research" (96.67 per cent) had received a greater response from farmers. Farmers felt that their innovations were exposed either little or large through PTD which otherwise would have gone unnoticed. This strong feeling and their commitment to work with scientists in a collaborative way had pushed them to give an overwhelming 'strongly agree' opinion towards this statement.

4.7. FACTOR LOADINGS OF PERCEPTION STATEMENTS AS PERCEIVED BY SCIENTISTS, EXTENSION PERSONNEL AND FARMERS

4.7.1. Factor Loadings of Perception Statements as Perceived by Scientists and Extension Personnel.

Among many perception statements selected for the study, some would have influenced PTD with greater prosperity than others. There is a possibility of some statements in conjunction with others would influence the PTD. In this context, Factor analysis would give the underlying pattern of relationship among the statements related to PTD. Through factor analysis the statements are converged into number of factors. The rationale behind this is to identify sum total of most influencing group of predictor statements on PTD.

The factor analysis was run in SPSS (Statistical Package for Social Sciences) using principle component factoring with varimax rotation as these measures had unique advantage of simplifying interpretation by maximizing the variances of the variable loadings on each factor.

As the same statements were administered to scientists and extension personnel, one factor analysis was run for both the respondents. The factor analysis has yielded five factors based on

Eigen values and percentage of variance accounted for each factor. The results are presented in Table 23.

Table 23. Eigen values and percentage of variance for each factor for Perception towards Participatory Technology Development (PTD) by the Extensionists and Scientists

Factors	Eigen values	Percentage of variance	Cumulative per cent of variance
Factor I	7.346	30.610	30.610
Factor II	3.473	14.469	45.080
Factor III	2.651	11.044	56.124
Factor IV	1.837	7.656	63.779
Factor V	1.614	6.725	70.504

The components having Eigen values greater than '1'were converged as factors. The first factor accounted for 30.61% of the variance followed by other factors accounting for 14.47%, 11.04%, 7.66% and 6.73% respectively. Together, these five factors accounted for 70.50% of variation in the perception of PTD by the Extensionists and Scientists.

The statements with larger factor loadings under each factor were presented in Table 24.

Table 24. Statements with factor loadings for Perception towards Participatory Technology Development (PTD) by the Extensionists and Scientists

Statements	Factor loadings
Factor I	
Look down on farmers' indigenous knowledge.	0.912
First hand appreciation of the diversity of farmers' problems	0.906
Help extensionists to recognize their roles	0.896
Link up with local problem – solving initiatives.	0.731
Researchers/Extensionists to interact out side the institution	0.701
Farmer Participatory Research	0.694
Collaborative activity and synergies between farmers, development partners and researchers	0.681
Generation of Win – Win technologies	0.635
Factor II	
Make research more responsive to farmer needs	-0.836

Incorporation of farmers' criteria into technology design	0.743
Involve farmers as co – investigators and researchers	0.682
Expand their imagination about different modes of research	0.644
Researchers need to have the ability to engage in dialogue, and listening	0.568
Factor III	
Organizing farmers into groups	0.695
Identification and use of ITK	0.575
Supporting farmers in learning by doing without involving researchers	0.559
Factor IV	
Knowledge is generated both by researchers and farmers	0.780
Focusing on local knowledge and resources	0.780
Researchers to convince on the superiority of their knowledge as compared to that of farmers	-0.594
Factor V	
Multi disciplinary approach	0.807
Farmer participation ends with good research	0.784

There were eight statements having significant loadings on factor I. They were look down on farmers' indigenous knowledge (0.912), first hand appreciation of the diversity of farmers' problems (0.906), help extensionists to recognize their roles (0.896), link up with local problem – solving initiatives (0.731), researchers/extensionists to interact outside the institution (0.701), farmer participatory research (0.694), collaborative activity and synergies between farmers, development partners and researchers (0.681) and generation of Win – Win technologies (0.635) in the descending order of factor loadings. Since these statements related to the personal ability of scientists and extensionists to interact with farmers these may be named as **'Role Performance Factors'**. The grouping in Ist factor clearly reveals that the role of three important systems viz., Scientists, Farmers and Extensionists are to be coordinated to make PTD a successful concept. Farmers have to extend their innovative ideas, ITKs to set the ball rolling. Whereas the scientists have to recognize the fact that win – win technologies should be developed, keeping in mind the concerns of farmers and scientists themselves. So their task is cut out that they have to provide a platform to farmers to share their ideas in a participatory manner. Similarly the role of extension personnel to synergize other activities in line with farmer's needs and scientists requirements are also emphasized in the grouping.

Factor II consists of five statements with higher factor loadings. They make research more responsive to farmer needs (-0.836), incorporation of farmers' criteria into technology design (0.743), involve farmers as co – investigators and researchers (0.682), expand their imagination about different modes of research (0.644) and researchers need to have the ability to engage in dialogue, and listening (0.568). These statements directly related to the farmers needs and involvement towards research. So these may be named as **'Research Aptitude Factors'**. The statements were grouped in II nd factor vividly explains the research inputs responsive to every stake holder's needs. It upholds the researcher component in experimentation by clearly accentuating the mutual understanding of farmers and scientists in designing themes and procedures of research. This power relationship among these systems is to be leveraged to make PTD responsive to local needs.

Factor III had organizing farmers into groups (0.695), identification and use of ITK (0.575) and supporting farmers in learning by doing without involving researchers (0.559) with higher factor loadings. These statements emphasize the organization of groups and use of ITK. So these may be named as **'Capacitating Factor'**. The III factor loading explains that farmer's capacity to organize themselves and make use of rich indigenous practices are vital for sustaining the concept of PTD. From time immemorial, farmers learnt what ever may do in the field which will be long lasting in their mind rather than whatever ideas may receive through their fellow farmers. The extension concept of 'Learning by Doing' is glued to the concept of PTD. The experimentation in their own field make the farmers build their capacity.

Factor IV consisted of three statements with higher factor loadings. They were knowledge is generated both by researchers and farmers (0.780), focusing on local knowledge and resources (0.780) and Researchers to convince on the superiority of their knowledge as compared to that of farmers (-0.594). Since these statements were directly related to sharing of knowledge between farmers and researchers. These may be named as **'Cognitive Factors'**. Knowledge is the base for every research endeavour. The factors got loaded in IV th factor stand tall to uphold knowledge generation through PTD. The knowledge generation in terms of fixing the experiments in the light of past experience, laying down experiments plots, carrying out the experiments, documenting the results are integral part of PTD.

Factor V consisted of two statements with higher factor loadings. They were Multi disciplinary approach (0.807) and farmer participation ends with good research (0.7840. These statements may be named as **'Stakeholders Factors'**. This factoring of statements were depicted in Fig. 32. This figure clearly revealed how the variables were grouped based on underlying pattern of relationship. Research is an unending process since every facet of research is ever widening and no one can tell the research is complete by his research findings. Similarly every researcher belonging to varied disciplines has their own orientation towards a research protocol. In recent times, multi – disciplinary research or involving relevant scientists of many disciplines is gaining momentum. In PTD also, the multi – disciplinary thought is important, since any technology evolved through PTD should be thoroughly ground in experiences of scientists of various disciplines. Their participation in research design to results sharing is very important.

4.7.2. Factor Loadings of Perception Statements as Perceived by the Farmers

The perception statement administered among farmers to know their perceived opinion about PTD was subjected to Factor analysis to delineate any underlying pattern exists among the statements. The factor analysis has yielded five factors based on Eigen values and percentage of variance accounted for each factor. The results are presented in Table 25.

Table 25. Eigen values and percentage of variance for each factor for Perception towards Participatory Technology Development (PTD) by the Farmers

Factors	Eigen values	Percentage of variance	Cumulative per cent of variance
Factor I	11.258	35.182	35.182
Factor II	4.843	15.135	50.317
Factor III	3.984	12.450	62.767
Factor IV	2.812	8.788	71.555
Factor V	2.173	6.790	78.345

It was disclosed from the Table first factor accounted for 35.18% of the variance followed by other factors accounting for 15.14%, 12.45%, 8.79% and 6.79% respectively. Together, these five factors accounted for 78.35% of variation in the perception of PTD by the farmers.

The statements with larger factor loadings under each factor were presented in Table 26.

Table 26. Statements with factor loadings for Perception towards Participatory Technology Development (PTD) by the Farmers

Statements	Factor loadings
Factor I	
Innovations initiated by local people /scientists are entry points	0.850
Strengthens farmer's organizations	0.835
Farmer's innovative capacities are strengthened.	0.779
Help to bring stakeholders together	0.769
Effective communication with local groups	0.692
On going (but not endless) process	0.672
Depend on collaboration among farmers, field workers, researchers	0.619
Researchers and farmers each have to contribute knowledge	0.597
Expansion of spheres beyond agriculture into marketing	0.579
Act as a researcher in the field by the farmers	0.536
Factor II	
Suitable for resource poor households	0.902
Bringing together the knowledge and research capacities of the local farming communities	0.844
Cost effective approach	0.835
Considering the Agro – climatic conditions in the research	0.833
Farmer- to-farmer networking	0.603
Sustained even without external agencies	0.583
Factor III	
Information exchange between farmer and researchers	0.846
Emerge the farmer's local ideas in a successive way	0.790
Scientific rationales behind the ITK are realized by the farmers	0.624
Use of farmer to farmer extension	0.508
Factor IV	
Wider adoption of agricultural technologies	0.818
Experimenting locally with a variety of options	0.788
Strengthening local capacities to experiment	0.754
Involvement of farmers in all stages of the technology development process	0.572
Arouses the farmers' inherent curiosity and creativity	0.566
Involves the farmers in all stages of the development process	0.562
Factor V	
Learning among farmers, scientists and extension workers	0.964
Satisfaction among farmers that their views/ideas are taken into consideration for evolving a technology	0.894
Ability of farmers to experiment and solve problems on their own	0.687

There were ten statements having significant loadings in factor I. They were "innovations initiated by local people /scientists are entry points" (0.850), "strengthens farmer's organizations" (0.835), "farmer's innovative capacities are strengthened" (0.779), "help to bring stake-holders together" (0.769), "effective communication with local groups" (0.692), "ongoing (but not endless) process" (0.672), "depend on collaboration among farmers, field workers, researchers" (0.619), "researchers and farmers each have to contribute knowledge" (0.597), "expansion of spheres beyond agriculture into marketing" (0.579) and "act as a researcher in the field by the farmers" (0.536) in the descending order of magnitude of loadings. Since these statements related to innovative capacities and innovation initiation by the farmers these may be named as **'Innovation Institution factors'.** The factoring of perception statements in Factor I clearly depicts the imperative of innovation of farmers in PTD. PTD, basically sharpen the innovative capacity of farmers to do such process continuously. This innovation process is endless, the utility of which is enhanced when it get together all stakeholders in the innovation process. Further the loaded statements into this factor reveal that innovation are to be institutionalized through organization building and collaboration with scientists and extension personnel wherein communication will be effective.

Factor II consisted of six statements with higher factor loadings. They were "suitable for resource poor households" (0.902), "bringing together the knowledge and research capacities of the local farming communities" (0.844), "cost effective approach" (0.835), "considering the Agro – climatic conditions in the research" (0.833), "farmer- to-farmer networking" (0.603) and "Sustained even without external agencies" (0.583). As these statements directly related to the cost effectiveness and location specificity of the research, may be named as **'Sustainability Factors'**. Economic benefits should go hand in hand with social benefits. At the end of the day, how much a farmer earns matters than anything else. In this context, the statements got grouped into factor II, are obviously interrelated and represent sustainability of the system. The research output tend to diffuse with greater pace, if it is cost effective, location specific and compatible with prevalence bio – physical and agro ecological situations. Many agricultural technologies got faded from the minds of farmers since they lack above traits. Apart from soundness of technologies, the sustainable utility of innovations largely hinges on the networking of farmers and their word of mouth spread. Besides any new idea in the form of technology refined through PTD, will have location specificity, cost effectiveness embedded in it.

Factor III had "information exchange between farmer and researchers" (0.846), "emerge the farmer's local ideas in a successive way" (0.790), "scientific rationales behind the ITK are realized by the farmers" (0.624) and "use of farmer to farmer extension" (0.508) in decreasing order of factor loadings. These statements emphasized on the networking of farmer with their fellow farmers and scientists, hence may be named as **'Networking Factors'.** Networking of interested farmers and scientists is very important for a successful PTS cycle and subsequent technology generation. In this process, appreciating the indigenous practices of farmers by the scientists is a precarious step towards forgoing an alliance. Farmer to Farmer Extension has been proved to be an important mode of extension other than public extension of officials concerned. The utility of such a channel in promoting the awareness about PTD has yielded success as the researcher could see the enthusiasm among farmers to pass on the results of PTD.

Factor IV consisted of six statements with higher factor loadings. They were "wider adoption of agricultural technologies" (0.818), "experimenting locally with a variety of options" (0.788), "strengthening local capacities to experiment" (0.754), "involvement of farmers in all stages of the technology development process" (0.572), "arouse the farmers' inherent curiosity and creativity" (0.566) and "involves the farmers in all stages of the development process" (0.562). Since these statements were directly related to involvement of farmers in experimenting and adoption of technologies, these statements may be named as **'Experimental capacity factors'.** Most of the statements in Factor IV represents PTD in terms of experimentation capacity of farmers, their role in all stages of experimentation and their creativity in designing experiments. As joint experimentation is the crux of PTD, there is no doubt these statements carrying one or other idea about experimentation in PTD would factored together. The more the inquisitiveness and experimental ability of farmers, the more will be their involvement in PTD process. As the process goes on, the farmer's capacity get improved and their partnership with scientists get strengthened.

Factor V consisted of three statements with higher factor loadings. They were "learning among farmers, scientists and extension workers" (0.964), "satisfaction among farmers that their views/ideas are taken into consideration for evolving a technology" (0.894) and "ability of farmers to experiment and solve problems on their own" (0.687). As these factors involves learning, satisfaction out of PTD, may be named as **'Mutual learning Factors'.** One important objective of PTD is cross learning among scientists, farmers and extension personnel, which ultimately results in abiding got improved among farmers. This culminates in satisfaction of farmers that their voices are heard and their ideas are given due recognition among scientific community. This mutual learning propels the stakeholders of PTD to sustain the concept.

4.8. MULTIVARIATE ANALYSIS OF VARIANCE (MANOVA) AMONG THREE CATEGORIES OF RESPONDENTS ON FOUR ATTRIBUTES

Multivariate Analysis of Variance was carried out to test the differences between the selected categories of the respondents on several dependent measures simultaneously. While ANOVA examines differences for a single dependent measure, MANOVA tests for differences on a set of dependent variables, provided the dependent variables are typically related. According to Larry (1991) and John (1991) MANOVA is typically performed for two reasons (i.e) statistical power and to control type I error. In the present research the Degree of dependability, Status of adoption, Assignment of weightage score and Evaluative perception of treatments in Banana are the dependent variables studied. Which were studied among three types of respondents namely Scientists, Farmers and Extension Personnel? The difference among four categories and these dependent measures were analysed through MANOVA and the results are presented in Table 27.

Table 27. Multivariate Analysis of Variance (MANOVA) showing significant difference in the Degree of dependability, Status of adoption, Assignment of weightage score and Evaluative perception of treatments in Banana

Multivariate Test			
Measure of Multivariate Test	**Value**	**F**	**Error df**
Wilk's Lambda	0.401	4.337**	60.000
Hotellings Trace	1.377	4.990**	58.000
Roy's largest root	1.288	9.983**	31.000

In SPSS (Statistical Package for Social Sciences) there were four test statistics for multivariate analysis of variance. But when the variables are highly related three test statistics namely Wilk's Lambda, Hotellings Trace and Roy's largest root were chosen.

Table 28. Test of between subjects effects for the Degree of dependability, Status of adoption, Assignment of weightage score and Evaluative perception of treatments in Banana

S.N	Dependent variables	Type III sum of Squares	Mean Square	F Value
1	Degree of dependability on treatments in Banana	412.167	206.083	1.058
2	Status of adoption of treatments in Banana	5670.389	2835.194	5.089
3	Assignment of weightage score to treatments in Banana	3270.500	1635.250	0.416
4	Evaluative perception of treatments in Banana	607.056	303.528	0.387

1. R Squared = 0.060; 2. R Squared = 0.236; 3. R Squared = 0.025;
4. R Squared = 0.023

From Table 28 it is clear that 'F' value for Wilk's Lambda, Hotellings Trace and Roy's largest root were significant at 1% level of probability. This indicated that the overall multivariate model had significant differences among the dependent variables studied. Given this overall idea, the interest will be to know the individual dependent variables status in the multivariate model. This forms second part of MANOVA and given in Table 28.From the table it could be interpreted that 'F' value for all the dependent variable studied were significant at 1 per cent level of probability which indicated that the three category of respondents significantly differ in terms of Degree of dependability, Status of adoption, Assignment of weightage score and Evaluative perception of treatments in Banana. The 'F' value and R squared value were more for Status of adoption of treatments in Banana and Degree of dependability in treatments in Banana then others. Finally the 'R' squared values for each variable appear in the foot notes. The third part of MANOVA deals with how different categories of respondents differ from each other on estimated marginal means of dependent variable. This was analysed and presented in Table 29.

From the table it could be seen that the estimated marginal mean was higher for farmers than Extensionists and Scientists with regard to all the dependent variables like Degree of dependability, Status of adoption, Assignment of weightage score and Evaluative perception of treatments in Banana. This shows that farmers are consistently ahead in terms of their faith and conviction on PTD than scientists and Extension Personnel. The marginal mean are approximately closer to each other for scientists and Extension Personnel with respect to degree of dependability and evaluative perception where as appreciable gap was found for status of adoption and assignment of weightage. Scientists are actually practicing the concept of PTD as they are one of the main stakeholders, they were ahead of Extension Personnel who confines to facilitation role only. Moreover it was stated elsewhere in this report that Extension Personnel did have a sense of disbelief which was exhibited in some of their opinion.

Table 29. Estimated marginal means of categories for dependent variables of Degree of dependability, Status of adoption, Assignment of weightage score and Evaluative perception of treatments in Banana

Dependent variable	Categories	Estimated Marginal Mean	Standard Error
Degree of dependability in treatments in Banana	Scientists	64.000	4.030
	Extensionists	65.833	4.030
	Farmers	71.917	4.030
Status of adoption of treatments in Banana	Scientists	98.583	6.814
	Extensionists	80.000	6.814
	Farmers	110.500	6.814
Assignment of weightage score to treatments in Banana	Scientists	173.167	18.088
	Extensionists	190.417	18.088
	Farmers	195.417	18.088
Evaluative perception of treatments in Banana	Scientists	111.917	8.080
	Extensionists	118.333	8.080
	Farmers	121.833	8.080

4.9. RESULTS OF THE TREATMENT COMBINATIONS EXPERIENCED BY THE PRACTISED FARMERS

4.9.1. Feedback from the practiced Farmer at K. Dharumathupatti

I am Arumugam living in K. Dharumathupatti. I am a full time farmer. I have been cultivating a variety of crops in my fields including Banana. I planted the poovan variety in my field.

Banana plantations in my area have not been considered much because of the problems faced by the farmers in cultivation and marketing. Previously the banana growers met with huge loss due to lack of shining, uneven size of the fingers getting a low price in the markets.

In this year 2008 at the time of flowering, Mr. Israel Thomas, a Ph.D Researcher, enlightened us in using farmers practices and scientific techniques (Indigenous Technical Knowledge) in banana. Myself and some of my fellow farmers applied farmer's practices and scientific techniques in the cultivation of banana in our fields as per the instructions given by him.

I followed farmer's practices and scientific techniques as a part of my banana plantation and cultivated the rest in the normal way. As days passed on, I have been able to see the difference between the bunches in the banana trees in which I followed organic farming techniques and the trees in which I have not. First of all, the fruits are shiny and the rib lines are prominent. They showed uniform growth of bunches from top to bottom, and even in size and girth. The average length of fingers has been significantly increased after the treatments. A glance at my field itself showed to others that there is some difference between untreated and others. We used 12 treatment combinations ranging from T1 to T12. Each treatment had a different combination than the other. Out of the 12 combinations the treatment combinations like T7 – application of fresh cow dung extract, zinc sulphate and borax, T8 - application of fresh cow dung extract with zinc sulphate and T9 - application of fresh cow dung extract with borax, showed great difference than the rest. As a whole the plants in which the treatments applied showed much difference when compared with the trees in which these treatments have not been applied. The treatment combination in T7 is growing very fast and it came to harvest 10 -15 days advance of the normal time.

The fruits showed even growth from top to bottom, the bunch showed shiny colour and keeping quality have been increased, the length of the fingers also increased. This yield parameters paved a way to fetches high rate in the market to the fruits. People in the market showed much eager in buying the treated bunches from my field. And day by day further recurring enquiries on the techniques that I have been used in my field.

My fellow farmers became astonished to see the improvements in my banana field and they enquired on the techniques that lead to this dazzling improvement. To all of them I replied that it is nothing but the use of the treatment combinations made by Department of Agricultural Extension, Agricultural College and Research Institute, Madurai of Tamilnadu Agricultural University. I explained them the treatment combination in detail and I have even asked Mr. Israel

Thomas to conduct more and more sensitization programmes in the use of these treatment combinations and the net results so that many banana farmers like me get benefited by it.

I wish to continue these treatments in all my banana fields in future. I am much satisfied to see the improvement in the fruits and the increase in the amount of income. In the past we were toiling for hours in the fields but at the time of marketing we used to get a very low rate, all our toil resulting in vain. But this new technique is a boon to farmers who grow banana and I think it is good for all my fellow farmers to use these treatments in their fields so that they may also be getting good income for their toil.

4.9.2. Feedback from the practised Farmer at K. Dharumathupatti

My name is Kankeyan and I am living in K. Dharumathupatti. I belong to a farmer's family and I have been cultivating banana for years. I have planted Red Banana in my field. At the time of flowering, Mr. Israel Thomas, a Ph.D Researcher, enlightened us on using farmers practices and scientific techniques (Indigenous Technical Knowledge) in banana. I wished to try it in my fields and I applied farmer's practices and scientific techniques in the cultivation of banana in my field as per the instructions given by him.

In the past I used to apply chemical fertilizers to improve the quality of the fruit, to make the fruits even and shining, to make the fingers with good length and weight so that it might fetch a good income to me. Though I applied many combinations of fertilizers and various other methods recommended by so many companies and agencies I did not able to see any improvement.

The banana plants in my fields when affected by diseases, in the past, I tried so many measures to save them, but of no use. When some of my banana plants affected by disease I thought that it is going to be same story. But the application of these treatments has rejuvenated the affected plants.

I have seen it through my own eyes. The fruits became shiny, the plants became lively. I have never seen anything like this before - from top to bottom the fruits are amazingly even in size and girth. The average length of fingers was significantly increased after the treatments.

A variety of treatment combinations ranging from T1 to T12 have been experimented in my field. Out of the twelve, the treatment combinations like T10 – application of Urea, Sulphate of potash, water and Fresh cow dung, T8 - application of fresh cow dung extract with zinc sulphate and T4 - application of Panchakavia with Household wood ash showed great difference when compared with the rest. As a whole the plants in which the treatments applied showed much difference when compared with the plants which was not treated.

In the past, I faced so many problems in marketing the fruits, because of the uneven size and length, non appealing colour, people were not ready to offer the fruits a reasonable prize. But now because of these treatments, the fruits showed even growth from top to bottom, the bunch showed shining colour and good keeping quality, the length of the fingers are good and because of all these reasons the fruit fetches high rate in the market. People in the market showed much eager in buying the bunches.

The above treatment may help me in getting good yield and price thereby enabling me to pay all my debts and a frictionless life. The support rendered by the PhD scholar Mr. Israel Thomas and the officials from Agricultural College and Research Institute, Madurai and ensured that the treatment combinations are used in the fields without any flaws. The officials who visited the fields gave us so many recommendations and suggestions that helped us in brining a revolution in our fields.

I wish to continue these treatments in all my banana fields in future. I have been much satisfied to see the improvement in the fruits and the increase in the amount of income. So I think it is good for all my fellow farmers to use these treatments in their fields so that they may also get good income.

4.9.3. Feedback from the practised Farmer at K. Dharumathupatti

My name is Jeya and I am living in K. Dharumathupatti. I belong to a farmer's family and I have been cultivating banana for years. Now I have planted Robusta variety in my field.

Through a Research Programme offered by the Department of Agricultural Extension and Rural Sociology, Agricultural College and Research institute, Madurai, Mr. Israel Thomas, a PhD Researcher, informed about farmers practices and scientific techniques (Indigenous Technical Knowledge) in banana. He motivated us saying that the application of this technique we would be able to see great improvement in our banana fields. He further persuaded that the implementation of these techniques would give us quality banana bunches with even sized fruits from top to bottom, good average length of fingers and good market price. I became motivated and I applied farmers practices and scientific techniques in the cultivation of banana in my field as per the instructions given by him.

In the past I was able to get fruits of some average quality that were uneven in size. Even when the size was good, the weight was not as expected and this created so many problems in marketing the fruits. I have longing for a technique that could make my banana plants lively and of good quality bearing shining and quality, even sized, lengthy fruits of some good weight.

The use of these new techniques worked well in my field. I was able to see it through my own eyes. The fruits became shiny with increased size, increased number of fruits, the tree became lively. I have never seen anything like this before - from top to bottom the fruits are amazingly even in size and girth. The average length of fingers has been significantly increased after the treatments. On the whole a drastic change appeared on my field giving life to bunches and fruits.

Out of the 12 treatment combinations used in my field, the treatment combinations like T8 - application of fresh cow dung extract with zinc sulphate, T9-application of fresh cow dung extract with borax and T10 - – application of Urea, Sulphate of potash, water and Fresh cow dung showed much difference than the rest. As a whole the plants in which the treatments applied showed much difference when compared with those not applied.

Though there is time to cut bunches and think about the income by marketing them, the picture that I get from my banana field is promising. My field stand is good when compared with the other farmers. I hope to get a good yield and some great income in the next few days. And a good sign is that people have been enquiring about the reasons for such improvement in my fields and I hope that soon they too would be using these techniques for better results in their fields.

4.9.4. Feedback by the practised Farmer at Kondayampatti

My name is Ramesh and I am a farmer of kondayampatti. My family has been in the occupation of cultivation for more than 20 years. I had planted banana (Grand – 9 Variety) this time and Mr. Israel Thomas, a PhD Researcher, sensitized us in using farmers practices and scientific techniques (Indigenous Technical Knowledge) in banana. I thought it to be a nice technique and so I implemented it in my field as per the guidelines given by him.

As days passed by I was able to see the improvement by my own eyes. From top to bottom the fruits grow amazingly even in size and girth. The average length of fingers has been significantly increased after the treatments. The trees became lively and the fruits showed good colour. The rib of the fruit is so thick. The fruits become shining, the fruits are even in size and the length of the fruits has been good.

Among the variety of treatments used in my field, the treatment combinations like T8 - application of fresh cow dung extract with zinc sulphate, T9 - application of fresh cow dung extract with borax and T10 - – application of Urea, Sulphate of potash, water and Fresh cow dung showed great difference when compared with the rest. As a whole the plants in which the treatments have been applied show much difference when compared with the plants in which these treatments have not been applied.

In the past, I have faced so many problems in marketing the fruits. Because of the uneven size and length, diseased looking, non lively colour, people were not ready to offer the fruits for a reasonable prize. But now because of these treatments, the fruits showed even growth, the bunch showed shining colour and utmost keeping quality, the length of the fingers were good and this enabled fruit fetches high rate in the market. People in the market showed much eager in buying the bunches. People said that the fruits from my field have been placed with the best fruits in the market. They enquired me on the techniques that I used in my field to get such quality fruits and I happily tell them about the treatments.

We got a good support from the research scholar Mr. Israel Thomas and various other officials from Agricultural College and Research Institute, Madurai of Tamil Nadu Agricultural University. They sensitized on the treatment combinations and they were with us in each and every step. They explained things very neatly and whenever we had doubts they cleared all our doubts without any hesitation. It is they who convinced us to go for the implementation of such new techniques in our field that has helped us in achieving such good yield and improvement in our fields.

These treatments have come to me as a boon and I wish to continue these treatments in all my banana fields in future. I have been much satisfied to see the improvement in the fruits and the increase in the amount of income. So I think it is good for all my fellow farmers to use these treatments in their fields so that they may get good income for their toil.

4.10. THE RESEARCHER'S SELF REFLECTION ON THE PARTICIPATORY TECHNOLOGY DEVELOPMENT (PTD)

Learning from the experience derived from the Participatory Technology Development (PTD), the researcher felt that self answering to a series of questions related to the PTD process in conducting experiments would be worthwhile. The following check lists (Table..) depicts the researcher's self experience starting from the conception, moving through the conduct and doing of PTD through field experiments, follow up and statistically and mentally recapitulating the results, reflections, farmers feedback and the 'net effect' of the PTD as a whole (i.e.) the sequential steps leading to the end ('the several parts making the whole').

Table 30. Check list of the researcher's self reflections on PTD.

Sl. No.	Statements	Strongly agree	Agree	Dis agree
1	The research addressed a concrete issue and practical problem	√		
2	There was participation by the farmers in all stages of the research programme	√		
3	The grounds for the partnership between farmers and researcher were explicitly negotiated and agreed	√		
4	The research was part of a continuous cycle of development rather than one – off project	√		
5	There was a clear view of how the research findings will feed back directly into practice	√		
6	The kind of PTD research used was clear to the researcher	√		
7	Experimenters and Farmers knowledge was acknowledged as a advantageous for the research	√		
8	The research was sufficiently small scale to be combined with a routine work load		√	
9	Ethical matters were taken into consideration		√	
10	The research provided solutions to a real problem/development	√		
11	The research went beyond a mere exploration of short – term solution	√		

12	The research acted as a base for social change	√		
13	The research treated the farmers conditions as a whole	√		
14	The research out put projected the satisfaction level of the banana growers	√		
15	The research has enriched academic and organizational initiatives		√	
16	The research has valued farmers knowledge and their participation in accepting the innovative things	√		
17	The research brought recognizable success quickly	√		
18	The research fit to local farming patterns and conditions	√		
19	The research was free of risk			√
20	The research has understood how the farmers understand and deal with the innovative practices		√	
21	The research was applicable and acceptable to all category of farmers	√		
22	The research was simple and understandable with respect to the farmers	√		
23	The research required only a minimum of on – site supervision			√
24	The research aroused enthusiasm and self confidence among farmers	√		
25	There was a feeling of optimism among the farmers throughout the programmes	√		
26	The research brought about enhancement in awareness, knowledge, skill, attitude and physical achievements	√		
27	The positive socio – psychological changes in the farmer's side were clearly observable during the research programme.	√		
28	The purpose and spirit of collaborative and participative mode was evident throughout the research	√		
29	The research resulted in the success of conducting experiments through PTD mode	√		
30	Farmers internalized with the concept of group spirit and Participatory approach		√	

The refinement of Indigenous Technical Knowledge through PTD is the main crux of the research. The researcher was aiming at uniform growth of bunches in Banana by conducting field experiments in the farmer's field by involving farmers in all the stages of PTD process. There was full participation from all the experimenter farmers and the participants. The farmers were involved in assessing importance of the research topic, fixing the treatment combinations, conducting experiments, application of the treatments in their experimented field, recording the observation, feedback and interpretation of results. Most of the time the researcher was a facilitator. The PTD process were agreed upon by the farmers and implemented. The technology identified will be logically and practically fitted into to the research work. The experimented farmers expressed their outstanding performance of the results of the research in their feedback and assured to do the same treatments in the next season crop on a continuous cycle. The sustainability of the technology was evident from their feedback workshop. The theoretical orientation and methodology were designed in such a way that the researcher had full comprehension about the Participatory Technology Development (PTD).

Though the research programme was more than the work load anticipated for a doctoral research and the researcher thoroughly enjoyed taking up such a challenge. In this research farmers socio – cultural economic and agro ecosystem situations were taken into consideration. The research was definitely a indispensable asset to the banana growers for getting uniform growth of bunches from top to bottom thus increases the yield and income level in the state of attractive and sustainable. There was an observable positive directed change in the experimented farmers and the participants of this research in terms of attitude, mindset and confidence in adoption of the tested technology. In this research farmers were done all the operations of the treatments so naturally they gained knowledge about doing things correctly and gained the confidence of how to do it in the future. Wherever similar situations occur this research results could be extrapolated.

Within the short span of time the researcher brought the recognizable success in Banana for the well being of the banana growers. The research was implemented by involving the banana growers in all the stages of PTD process and experiments stages. The farmers have out lined and implemented the treatment combinations. The research aroused enthusiasm and confidence among the banana growers. The skepticism and pessimism among the farmers changed to optimism, both quantitatively and qualitatively. The technology tested in the field got a very good appreciation among the farmers, scientists and the extensionists. These technologies will be highly useful to the banana growers in a longer run to get uniform growth of bunch to ensure more yield and income. The tested technologies are highly rationalized by the scientists and extensionists so that the adoption rate will be higher in the near future.

SUMMARY AND CONCLUSION

India over several millenniums has been the treasure land of biological wealth, intellectual knowledge and spiritual wisdom. During both pre and post Vedic periods, arts, science and technology had flourished well in various fields. Indian agriculture to some extent is much traditional and dates back to about 10,000 years. India with its traditional crop varieties and crop production technologies was able to feed its population and also exporting some selected produces to other continents. The development of modern agricultural production was achieved by creating large scale specialized farm production units, increased mechanization and use of chemical inputs.

Apart from the substantial increase in the crop yields through green revolution, these modern technologies have made the Indian farmers become poorer as they required heavy investment on costly external resources of uncertain future availability. These technologies with the principle of monoculture resulted in the rapid erosion of crop and livestock diversity, natural soil fertility and biological pest regulation, enhanced the soil erosion, salinization and environmental problems, reduced the nutrition once obtained from the wild foods on farms, increased the need for expensive and poisonous chemical fertilizers and pesticides and finally made the farmers more dependent on markets and outside agencies resulting in an imbalanced growth between rural and urban areas. The impact was more on millions of small farmers and pastoralists. Hence, a section of policy makers and agricultural scientists have now understood that continuation of modern agriculture might lead to severe ecological and economic problems. They are also convinced that modern agriculture may not be able to meet the requirements of the ever increasing population in the future. So, they are searching for alternative technologies. Several alternatives have been proposed such as low external input agriculture, sustainable agriculture, organic farming, biodynamic farming etc. However, they require sometimes little or considerable external inputs whose availability may be uncertain in future.

In developing countries, the other alternatives viz., traditional methods have special advantages over modern agricultural techniques. The capital and technological skill requirements in the use of traditional technologies are generally low and their adoption often requires little restructure of the traditional societies. These traditional technologies are not other than the indigenous knowledge. By adopting such indigenous knowledge, our ancestors did not face any problem of large scale pest out break or economic crisis unlike the today's farmers.

Indigenous knowledge system constitutes an alternative model for sustainable development. It is the sum total of experience and knowledge of a given group that can be used to form the basis for decision-making in the face of familiar and unfamiliar problems and challenges. One way to narrow the gap between local communities and outside professionals is by getting closer to local people and learning from them. By getting closer is meant establishing better rapport and improved channels of communications. Farmer Participatory Approach helps in interacting with local communities, understanding them and learning from them. It helps in the process of involvement with local communities for indigenous knowledge building exercises. It is a way of learning from and with community members to investigate, analyze and evaluate constraints and opportunities and make informed and timely decisions regarding development projects.

Participatory approaches emerged as a way to address the technology needs of highly diverse farming systems in marginal environments, where resources are limited and conventional approaches to research and development often failed. Participatory approaches range from researcher-driven studies that simply involve farmers, to project in which farmers set the agenda, empowerment and education are key issues, and the participation of diverse stakeholders is sought. "Participation" in this effort includes notable cooperation among researchers, extension workers, private machine shop owners, credit institutions, non-government organizations, and others.

Having this scenario, in the back drop that the present study was taken up with the following objectives.

1. To study the Socio – Personal and Psychological characteristics of the farmers in the study area.
2. To identify the Indigenous Agricultural practices amenable for Participatory Technology Development (PTD).
3. To operationalize the steps involved in PTD Process and build the platform for Joint experimentation for farmers and Researchers.

4. To conduct experiments to refine the identified Indigenous Agricultural Technology for better adoption.
5. To compare the refined Indigenous Agricultural Technology with the scientific one.
6. To assess the usefulness of the technologies developed and the scope for up scaling the PTD process.
7. To test the refined technologies for successful advocation.

Phase I of the Study

The field experiments were conducted during January to May 2008 at K. Dharmathupatti village of Reddiarchathiram Block, Dindigul District and Kondayampatti village of Alanganallur block, Madurai District of Tamil Nadu which were selected purposively for conducting the field experiments to study the influence of post - shooting application of various ingredients to enhance uniform growth of bunch from top to bottom in Banana.

Key Informants workshop was conducted with the progressive banana farmers and the scientists in the Banana belts of K. Dharmathupatti, Kondayampatti and Melakkal to finalize the treatments with farmer participatory approach. G-9, Robusta, Poovan and Red banana varieties of Banana crop were chosen for the experiments in the study. The following 12 treatments were finalized at the end of the key informant's workshop.

The treatments were

T_1 - House hold wood ash + Pregnant cow urine

T_2 - Brick Chamber wood ash + pregnant cow urine

T_3 - Panchakavia + Brick Chamber wood ash

T_4 - Panchakavia + Household wood ash

T_5 – Urea

T_6 - Urea + Sulphate of Potash

T_7 - Fresh cow dung extracts + Zinc sulphate + Borax

T_8 - Fresh cow dung extracts + Zinc sulphate

T_9 - Fresh cow dung extracts + Borax

T_{10} - Urea + Sulphate of potash + water + Fresh cow dung,

T_{11} - Cow dung extract solution

T_{12} – Control.

The treatments were used for post – shooting application and applied by the following way. The distal axis (male axis) was cut 15 cm away from the last hand fifteen days after the last hand was opened and after removal of the distal axis along with the navel bud (male bud). The ingredients of the treatments were tied in a polythene bag. Harvesting was done uniformly at full maturity stage by visual observation. Biometrical observations on the yield parameters under each treatment replication wise and of bunch were done. Data on the yield parameters were recorded at the stage of harvest. They are Yield (t/ha), Bunch Weight (kg), Number of hands and fingers, Average weight of the fruits (fingers), Volume of the fruit (finger), Density (Volume/weight ratio) of fruit (fingers), Average Length of the fruits (fingers), Girth of the Fruits (fingers), Total Soluble Solids (TSS) and Total Number of fingers/bunch.

PHASE II OF THE STUDY

Representatives from all the sub systems (stakeholders) namely Farmer Sub System (FSS), Extension Sub System (ESS) and Research Sub System (RSS) were selected with 30 respondents each in every sub system. Based on the objectives and by reviewing relevant literature and discussion with experts, the following broad attribute measures were selected for study: Degree of dependability in treatments in Banana, Status of adoption of treatments in Banana, Assignment of weightage score to treatments in Banana, Evaluative perception of treatments in Banana and Rationalization of Treatments in Banana. Besides Perception towards Participatory Technology Development (PTD) were elicited from all the three systems. The analysis was done by using Path Analysis, Kendall's Coefficient of Concordance (Kc), Spearman's Rank Order Correlation, Factor Analysis and Multivariate Analysis of Variance (MANOVA). The efficiency of the treatments were analysed using Simple Randomized Block Design.

The salient features of the study are furnished below.

RESULTS OF FIELD EXPERIMENTS

I. The results of treatments conducted at Kondayamapatti Village in Grand – 9 variety

1. The average length, average girth, average weight of fingers, number of finger's per hand, total number of fingers per bunch, No of hands/bunch, bunch weight, the yield of Banana and TSS was significantly increased in T_8 which is application

of fresh cow dung extracts and zinc sulphate. The average volume of finger was not influenced by the treatments, average density of the fingers found to be significantly higher in all the treatments

2. The higher positive and significant association was obtained between yield and bunch weight and Finger weight, followed by number of fingers per hand.
3. Treatment T_8 which is the application of fresh cow dung extracts and zinc sulphate resulted in the highest benefit cost ratio of 3.0 followed by 2.9 in T10 (application of urea, sulphate of potash, water and fresh cow dung).

II. The results of treatments conducted at K. Dharmathupatti Village in Robusta variety

1. The average length was significantly increased in treatment which is the T_4 application of panchakavia and household wood ash. The average girth was significantly increased in treatment T10 (i.e.) application of urea, sulphate of potash, water and fresh cow dung. The maximum fruit size, number of finger's per hand, total number of fingers per bunch, number hands per bunch, bunch weight, yield of Banana and TSS were observed due to the application of fresh cow dung extracts and zinc sulphate in T8. The average volume of finger was not affected by the treatments. The average density of the fingers found to be significantly higher in all the treatments.
2. The higher positive and significant association was obtained between yield and bunch weight, followed by the fingers per bunch, Volume of finger, Finger weight and number of hands per bunch.
3. The application of fresh cow dung extracts and zinc sulphate in T8 resulted in the highest benefit cost ratio of 2.2 followed by 2.1 in T1 and 2.1 in T9.

III. The results of treatments conducted at K. Dharmathupatti Village in Poovan variety

1. The average length and the number of fingers per hand were significantly increased in treatments T_8 (i.e.) application of fresh cow dung extracts and zinc sulphate. The average girth of fingers was significantly increased in treatments T4 (i.e.) the application of panchakavia and household wood ash. The maximum fruit size, density, bunch weight and yield of

Banana were increased due to the application of fresh cow dung extracts and borax in T_9. The average volume of fingers was not influenced by the treatments. The total number of fingers per bunch and TSS were significantly increased in treatment T_7 (i.e.) application of fresh cow dung extracts, zinc sulphate and borax. The spraying of panchakavia and brick chamber wood ash in T_3 recorded the highest number of hands per bunch.

2. The higher positive and significant association was obtained between yield and bunch weight, followed by Finger weight, Density of fingers, number of fingers per hand, number of fingers per bunch, and number of hands per bunch.
3. The application of fresh cow dung extracts and Borax in T9 resulted in the highest benefit cost ratio of 1.7 followed by 1.6 in T8

IV. The results of treatments conducted at K. Dharmathupatti village in Red Banana variety

1. The average length of fingers was significantly increased in treatment T_{10} (i.e.) application of urea, sulphate of potash, water and fresh cow dung. The average girth of fingers was significantly increased in T_3 which is application of panchakavia and brick chamber wood ash. The maximum fruit size, fingers volume, density of the fingers, number of fingers per hand, total number of fingers per bunch, number of hands per bunch, bunch weight, yield and TSS were observed in T_8 which involves application of fresh cow dung extracts and zinc sulphate.
2. The higher positive and significant association was obtained between yield and bunch weight, followed by Finger weight, number of fingers per hand, number of fingers per bunch, number of hands per bunch and density of fingers.
3. The application of fresh cow dung extracts and zinc sulphate in T_8 resulted in the highest benefit cost ratio of 3.7 followed by 3.6 in T10.

RESULTS OF PHASE II STUDY

1. Degree of dependability of treatments in Banana

The mean score obtained for degree of dependability of treatments revealed that the highest mean score for scientists was observed in

T10 (2.63) followed by T7 (2.53). The farmers expressed that the treatment T7 (2.90) showed the highest mean score followed by T10 (2.80). The highest mean score was recorded for extensionists in T10 (2.73) followed by T5 (2.67).

2. Status of Adoption of treatments in Banana

The results of status of adoption of treatments showed that the highest mean score as perceived by scientists was T5 (4.47) followed by T6 (4.30). The highest mean score was gained by farmers in T10 (4.07) followed by T11 (4.03). The highest mean score as perceived by the extensionists was T5 (4.13) followed by T6 (3.40).

3. Assignment of weightage score to treatments in Banana

The assignment of weightage score to treatments among scientists revealed that the highest mean score was obtained in T10 (8.70) followed by T8 (7.47). The highest mean score was attained by farmers in T10 (8.6) followed by T9 (8.33). The highest mean score to the treatments assigned by the extensionists was in T10 (8.170 followed by T5 (7.83).

4. Evaluative perception of treatments in banana

As far as the scores for evaluative perception is concerned, the highest mean score for scientists was observed in T10 (4.63) and T7 (4.33). The highest mean score as per farmers point of view was observed in T10 (4.8) followed by T8 and T9 (4.77). The highest mean score was observed by the extensionists in T10 (4.77) followed by T8 (4.4).

5. Rationalization of treatments in Banana

The highest means score for scientists was observed in T10 (4.47) followed by T7 (4.3). The highest mean score was obtained by Extensionists in T5 (4.57) followed by T10 (4.5).

6. Perception of Scientists towards to Participatory Technology Development (PTD)

More than two third (66.67%) of scientists agreed the perception statement "PTD is a collaborative activity involving farmers, scientists and extensionists for identification, appreciation, rectification of farmers innovative ideas". Nearly one fourth (26.67%) of scientists strongly agreed the statement.

7. Perception of Extension Personnel towards to Participatory Technology Development (PTD)

Cent per cent of extension personnel who were interviewed strongly agreed or agreed the perception statements. This shows the belief and commitment of Extension personnel towards PTD. Among the perception statements, "PTD is the incorporation of farmers knowledge into technology design and evaluation", "PTD is a multi disciplinary approach and communication between farmer to farmer, agency to farmers and scientists to farmers", and "Knowledge sharing between farmers and scientists/extensionists" were endorsed by strong agreement of 66.67 per cent, 60.00 per cent and 56.67 per cent of Extension personnel respectively.

8. Perception of Farmers towards to Participatory Technology Development (PTD)

Majority (more than 80 per cent) of respondents were fully aware of the benefits of PTD as a process of bringing agricultural science and scientist closer to them. All the perception statements received a "strongly agree" opinion from the respondents with a few opted for 'Agree' opinion. The statements namely "PTD is a knowledge contribution by scientists/extensionists and farmers" (93.33 per cent), "PTD is a farmer to farmer network" (90.00 per cent), and "PTD is satisfactory information exchange between farmers and researchers" (90.00 per cent) were having an edge over other statements in terms of degree of agreement as more number of respondents were in 'strongly agree' category than 'agree' category.

9. Factor Loadings of Perception Statements as Perceived by Scientists and Extension Personnel.

Five factors namely Role Performance Factors, Research aptitude factors, Capacitating Factor, Cognitive Factors and Stakeholders Factors emerged from factor analysis which explained 30.610, 14.469, 11.044, 7.656 and 6.725 percentage of variance in perception respectively. Totally 70.504 percentage of variance is explained by five factors.

10. Factor Loadings of Perception Statements as Perceived by the Farmers

Five factors namely Innovation Institution factors, Sustainability Factors, Networking Factors, Experimental capacity factors and Mutual learning Factors emerged from factor analysis which explained 35.182, 15.135, 12.450, 8.788 and 6.790 percentage of variance in perception respectively. Totally 78.345 percentage of variance is explained by five factors.

11. Multivariate Analysis of Variance (MANOVA) among three categories of respondents on Four Attributes

Multivariate Analysis of Variance (MANOVA) reveals that the estimated marginal mean was higher for farmers than Extensionists and Scientists with regard to all the dependent variables like Degree of dependability, Status of adoption, Assignment of weightage score and Evaluative perception of treatments in Banana and also indicated that the overall multivariate model had showing significant differences existed among different dependent variables selected for the study.

12. Comparative analysis of attribute measures of the treatments in banana among Research Support System (RSS), Farmers Support System (FSS) and Extension Support System (ESS)

a. Degree of dependability in treatments in Banana

There was no significant relationship between the attributes of Degree of dependability in treatments in Banana within FSS and ESS. There was positive and significant relationship between the attributes of Degree of dependability in treatments in Banana within FSS and RSS at one per cent level. There was positive and significant relationship between the attributes of Degree of dependability in treatments within RSS and ESS at five per cent level.

b. Status of adoption of treatments in Banana

There was no significant relationship between the attributes of status of adoption of treatments in Banana within FSS and ESS. There was positive and significant relationship between the attributes of status of adoption of treatments in Banana within FSS and RSS at five per cent level. There was positive and significant relationship between the attributes of status of adoption of treatments within RSS and ESS at one per cent level.

c. Assignment of weightage score to treatments in Banana

There was positive and significant relationship between the attributes of assignment of weightage score to treatments in Banana within FSS and ESS only at five per cent level. There was positive and significant relationship between the attributes of assignment of weightage Score to treatments in Banana within FSS and RSS at one per cent level. There was positive and significant relationship between the attributes of evaluative perception of the treatments within RSS and ESS at one per cent level.

d. Evaluative Perception of Treatments in Banana

There was positive and significant relationship between the attributes of evaluative perception of the treatments in banana within FSS and ESS at one per cent level. There was positive and significant relationship between the attributes of evaluative perception of the treatments in banana within FSS and RSS at one per cent level. There was positive and significant relationship between the attributes of evaluative perception of the treatments in banana within RSS and ESS at one per cent level.

e. Scientific rationality of treatments in Banana

There was positive and significant relationship between the attributes scientific rationality of treatments in banana with in RSS and ESS only at five per cent level.

Implications and Recommendations of the Study

Research Support System (RSS)

- The study revealed that application of fresh cow dung extracts and zinc sulphate in T8 and application of urea, sulphate of potash, water and fresh cow dung was found to be more effective in the farmer's field condition. The researchers should take cognizance of this results and think of up scaling the findings in other agro – climatical conditions prevailing in other districts through on – farm testing. After rigorous experimentation in various conditions. This may be revealed as a technology package which can be added into crop production guide for wider reach.
- The methodology requires scientists and extension personnel to work as equal partners with farmers in information sharing and implementation with regard to the treatments experimented in the farmer's field. Besides demonstrations of successful treatments on farmers fields, individual farmer experimental sites, joint researchers, farmers and extensionists research projects on farmers fields, outreach and farmer field days and farmers participation in stakeholders workshops are other tools that facilitates and fasten the

process of information exchange, technology development and testing, adoption and dissemination with the farmer at the steering seat. It has shown to be fast in technology development and dissemination to many farmers at a desirable rate. The approach is recommended for further testing before recommending its wide adoption.

- Researchers had always focused on biophysical aspects of the research process, but due to emphasis on the need to reflect on how the research process is affecting them, their research programs and interactions with colleagues and farmers have now been recognized as important aspects. In this regard, the scientists should give much importance to the farmer's needs and their innovative ideas into the research agenda for better participation farmers in the research aspects in a effective manner.

Extension Support System (ESS)

- The study revealed that the extension personnel were confident with IIHR and TNAU recommendations with regard to banana cultivation. The extension personnel should arrange training programmes, field days and campaign through Farmer Science Club (FSCs) to popularize the aforesaid recommendations besides the application of cow dung extracts and zinc sulphate.
- Workshop series and periodic review meetings have to be arranged to the team members involving scientists, farmers and extensionists for an opportunity to openly discuss the challenges in adapting the viable indigenous technologies, the participatory research process, modifications, and areas that require further capacity building and institutional support.
- Extension system can benefit from the realization that its work lends itself easily to the participatory processes of PTD and the exchange of scientific and indigenous knowledge from farmer-to-farmer. Keeping this in mind the extensionists can arrange a participatory meeting to encourage the farmers to come up with the indigenous knowledge in various fields and crops. Finally to elicited the viable practices to be implemented in that area as same as of this study.

Farmer Support System (FSS)

- The study revealed that the experimented farmers were quiet satisfied with the treatments in banana with regard to all the spheres. So they can make an initiative to arrange meetings at their respective village or block to recommend the best treatments suited to their area and field conditions for effective adoption.

- The important tool for participatory technology development and dissemination is participation of farmers in various stake-holder workshops. Their physical presence and participation in deliberations inspires them work more and much better than before. So FDGs can arrange the stake holder's work-shops with the involvement of Scientists and extension personnel to make further refinement of the well satisfied treatments for a longer run as a promising technology to their respective areas.
- Experimented farmers successful treatment combinations can be used as a technology in the demonstration sites to teach other farmers. Village members are informed of the day, time and venue for training. The training may be carried out by the experimented farmer at his/her farm. He can demonstrate the ways and means to carry out the treatments to other farmers and explains why the practice works. Interactions between the experimented farmer, other participating farmers, researchers and extension personnel continue during the demonstration. Where appropriate, researchers/extension chip in scientific facts that support performance of the practice and possibilities for improvement. Participating farmers may offer even better practical examples than those demonstrated by the expert trainer

SUGGESTIONS FOR FUTURE RESEARCH

Learning from the experience of the present study it is recommended that Participatory Technology Development of emancipation mode be institutionalized by local and regional research, extension and development systems of public, non – governmental and voluntary organizations.

Banana is an important crop and its utility was highly commendable. The refinement of Indigenous Technology in Banana was done with field experiments. It is also suggested that studies similar nature be conducted throughout Tamil Nadu, especially concentrating on other parts of Banana belt and also with other crops. Results of such technology assessment efforts may be quoted as examples during the transfer of technology endeavours.

REFERENCES

Abigail, Andah 2000. Technological Transitions: Technical upgrading of Indigenous Food Technologies in Africa. Food security and Sustainable Development Division (FSSDD) Economic Commission for Africa.

Adegunloye, D.V., Adetuyi, F.C., Akinyosoye, F.A. and M.O. Doyeni 2007. Microbial Analysis of Compost Using Cow dung as Booster, Pakistan Journal of Nutrition 6 (5): 506 – 510.

Agrawal, A. 1995. Dismantling the Divide between Indigenous and scientific knowledge. Development and Change 26 (3): 413 – 439.

Agrawal, A. 1995. Dismantling the Divide between Indigenous and Scientific Knowledge. Development and Change. 26: 413-439. In: Recording ND Using Indigenous Knowledge: A Manual (Ed.) International Institute of Rural Reconstruction, Silang, Cavite, Philippines. 1996: pp.7.

Ahamad, A.H., Sachan, A., Sharma, L.D. and Hore, S.K. 2002. Ethno Veterinary Medicine: Proclaiming its Resurgence, Indian Fmg, 35 (1): 35-38

Ahamed, P., Rajendran, P., Ramesan, K.K. and Potti, N.N. 1996. Farmer's criteria for selecting rice varieties give new sight to rice breeders. National Symposium on Technology Advancement in Rice Production, 17-19 June 1996. Directorate of Extension, Kerala Agricultural University, Thrissur, *Abstract*: 209.

Ahmed, M.M. 1994. Indigenous Farming Systems, knowledge and practices in Sudan, University of Khartoum, Sudan. 320 pp.

Alamelu, 1996. Technical Knowledge in Rural Farmers Kisan World, 23(10): 27-28.

Alcorn. J.B. 1984. Huastec Mayam Ethnobotany. 2nd ed. University of Texas Press, Austin. P. 176.

Altieri, M. 1991. Why Study Traditional Agriculture? In: Agro ecology (Eds.) C. R. Carroll *et al.,* pp. 551-564.

Altieri, M. 1996. Indigenous Knowledge – revalued in Andean agriculture. ILEIA News letter, 12 (1): 7-8.

Altieri, M.A. and M.K. Anderson 1986. An Ecological Basis for the Development of Alternative Agricultural Systems for Small Farmers in the Third World. American Journal of Alternative Agriculture, 11 (1): 20-28.

Amanor, K.S. 1991. Managing the Fallow: Weeding Technology and Environmental Knowledge in the Kerobe District of Khana. Agriculture and Human Values. 8: 5-13.

Andah, A. and Shoemaker A. 1994. Gender Aspects in Post – Harvest Technology A working paper prepared for the workshop on "Experiences gained in Africa on the improvement of Post–Harvest Technologies", July 4 – 8, 1994, Accra

Andrew, P. 1978. Technology and Peasant Production: Reflections on a Global Study, In: International Perspectives in Rural Sociology (Ed.) Howard. Pp. 183 -211. John Wiley and Sons, New York.

Arulraj, S. and J. Vasanthakumar 1996. "Participatory Technology Development – A Case study'. Discussion Paper 96/1. Sugarcane Breeding Institute, Coimbatore.

Ashby, J., C.A. Quiros and Y.M. Rivera 1987. Farmer Participation in on – farm varietal trials. Agricultural Administration (Research and Extension Net work). Discussion Paper No. 22. London: Overseas Development Institute.

Atte, D.O. 1989. Indigenous Local Knowledge as a Key to Local Level Development: Possibilities, Constraints and Planning Issues in Studies in Technology and Planning Issues in the Context of Africa. Seminar on Reviving Self-reliance, 3-5 February, 1989. United Nations Centre for Rural Development, Nagoya. Abstract: 13.

Babu, M.N. 1995. Evaluative Perception of Homestead Farmers in Relation to Appropriateness of Framing Systems and Cropping Patterns. Un published M.Sc (Ag.) Thesis, Kerala Agricultural University, Thrissur, India, p. 128.

Babu, M.N. 2000. Learning from traditional farming. Kissan Wld. 27(7):11-12.

Balasubramanian, P. 1992. Indigenous Knowledge Use in Dry lands: An Exploratory study. Un Published M.Sc. (Ag.) thesis, Tamil Nadu Agricultural University, Coimbatore. P. 169.

Baldeo Singh, R.N. Padaria, B.M. Sharma, M. Pathak and Jitendra Kumar 2007. Participatory Validation of Indigenous Agricultural Technologies and Farmers Experimentation Capacity Building. Indian Farming, Vol. 57 (4): 11 -14.p.

Bardhan, P. 1989. Conversations between Economists and Anthropologists. Oxford University Press. London, United Kingdom, 65 p.

Bentley, J.W. 1989. What farmers don't know can't help them; the strengths and weaknesses of Indigenous Technical Knowledge in Honduras. Agriculture and Human Values 6, PP. 25-31.

Bentley, J.W., Rodriguez, G. and Gonzalez, A. 1994. Science and People: Honduran Oampesinos and natural pest control inventions. Agriculture and Human Values, 11(2 & 3): 178 -182.

Benz, B.F., Cevallos, E.J., Santana, M.F., Rosales, A.J., and Graf, M.S. 2002. Losing Knowledge about Plant use in the Sierra De Manantlan Biosphere Reverse. Econ. Bot. 54 (2): 18-19.

Berkes, Fikert 1999. Role and Significance of "Tradition" in Indigenous Knowledge. Indig. Know & Dev. Monitor. 7 (1): 19.

Berkes, Fikert, Folke, Carl and Gadgil, Madhav. 1993. Traditional Ecological Knowledge, Biodiversity, Resilience and Sustainability. Stockholm, Sweden, Beijer International Institute of Ecological Economics. 34 pp. Beijer Discussion Paper Series No. 31.

Bharara, L.P. and Seeland, K. 1993. Indigenous knowledge and drought in the arid zone of Rajasthan, India: weather prediction as a means to cope with hazardous climate. International reihe No. 93/3.Department of Wald – Und Holzforsching-eth Zurich, Zurich. P. 17.

Biggelaar, C.D. 1991. Farming systems development: Synthesizing Indigenous and Scientific Knowledge Systems. Agriculture and Human Values 8(1/2): 25 -36.

Biggs, S.D. 1989. Resource Poor Farmer Participation in Research: A Synthesis of Experiences from Nine Agricultural Research Systems. Comparative Study. Paper No. 3. Overseas Development–Institute, London, 85 p.

Biggs, Huntley and H. Tinnermeir 1974. Small Farm Agricultural Development Problems, Colorado State University, Fort Collins, Colorado.

Breemer, J.P.M. 1989 Yam cultivation and Socio- Ecological Ideas in Aouan Society, Ivory Coast: A Constitution to Crop Sociology, Sociologia Ruralis, 29: 265-279.

Brokensha, D.W., Warren D.M. and Werner, O. 1980. Indigenous Knowledge Systems and Development. Lanham, MD: University Press of America.

Brouwers, J.H.A.M. 1993. Rural people's response to soil fertility decline, the Adja case (Benin). Wageningen Agricultural University Papers 93-4.

Brouwer, J. 1998. IK, IKs and ITK in IK: The Proof is in the Eating of the Pudding. Indig. Know & Dev. Monitor. 6(3): 13.

Careasco, E., Estrado, N., Gabriel, J., Aslfaro, G., Larondelle, Y., Garcia, W. and Quiroga, O. 1997. Seis cultivars potenciates delupa con resistancia altrizon tardio (*Phytophthora infestans*) en Bilivia. Revista Lationamerica dela papa 9: 106 -122.

Carter, H.O. 1988. The Agricultural Sustainability Issue: An Overview and Research Assessment. In: The Changing Dynamics of Global Agriculture. A. Seminar/Workshop on Research Policy Implications for NARS, DSE/ Zelfeldfing, Germany.

Chakravarthy, K. 1982. Indigenous Farm Practices: their Influence. Un published M.Sc. (Ag) thesis, Tamil Nadu Agricultural University, Coimbatore, p.132.

Chaman Lal and L.R. Verma 2006. Use of certain Bio – Products for Pest control, Indian Journal of Traditional Knowledge, 5(1): 79-82.

Chambers, R. 1988. Sustainable Rural Liveli hoods: A key strategy for people, Environment and Development. In the Greening of Aid: Sustainable Livelihood in practice, Earthscan Publications, London. Pp. 1-17.

Chambers, R. 1991. Shortcut and participatory methods for projects. Putting people first: Sociological Variables in Rural Development (ed. Cemea, M.M.), Oxford University Press, London, pp.31-76.

Champion, J., F. Dugain, R, Maignien and Y. Dommergues 1958. Les Sols De Bananeraies et leura melioration en Guinee, Fruits, 13: 415-462. Fruit Nutrition, Childers, N.F. (Ed.), Horti. Publns., New Jersey, USA, pp 52-77, 1966.

Chande, S. 1993. "Developing a Strategy for Integrating Indigenous Knowledge System in the Area of Foods, Nutrition and Family Welfare with formal Research System". Paper Presented at the National Seminar on Indigenous Technologies for Sustainable Agriculture, New Delhi, March. 23-25.

Chandra. D., Ghorai, A.K., Khan, A.R. and Nanda, P. 2000. On-farm Evaluation of Straw mulch on pointed gourd Production. Indian Farming, 50 (7): 37-40.

Chitamber, J.B. 1961. Extension Education in Community Development. Directorate of Extension, New Delhi, pp. 148-149.

Chittiraichelvan, R. and K.V. Raman 1990. Indigenous Knowledge of Farmers: Its Use in Extension Strategies for Rainfed Agriculture. In: Technologies for Minimizing Risk in Rainfed Agriculture (Eds.) S.P. Singh and C. Prasad, ISEE and ICAR, New Delhi.

Chittiraichelvan, R. and K.V. Raman 1991. Indigenous Knowledge of Farmers: Its Use in Extension Strategies for Rainfed Agriculture. International Conference on Extension Strategy for Minimizing Risk in Rainfed Agriculture, 17-19 March, 1991, Indian Society of Extension Education, New Delhi, Abstract: 69.

Compton, L.J. 1989. Strategies and Methods for the Access, Integration and Utilization of Indigenous Knowledge in Agriculture and Rural Development. In: Indigenous Knowledge Systems: Implications for Agriculture and International Development (Eds) D.M. Warren *et al.*, pp. 116-120. Iowa State University No. 11.

De Boef and W.K. Wellard 1993. Cultivating knowledge: Genetic Diversity, Farmer Experimentation and crop Research, London: Intermediate Technology Publications.

De Walt, B.R. 1994. Using Indigenous Knowledge to Improve Agriculture and Natural Resource Management, Centre for Latin American Studies, University of Pittsburgh, USA, p. 349.

Dommer, A.J. 1988. Innovation in African Agriculture, West View Press, Boulder.

Drink Water, M. 1992. The State and Agrarian Change in Zimbabwe's Communal Areas, Routledges, London.

Dubey, V.K., G.S. Naraina and S.L. Gupta 1993. "Methodologies for Tapping and Documenting Indigenous Technologies." Paper Presented at the National Seminar on Indigenous Technologies for Sustainable Agriculture, New Delhi, March. 23-25, 1993.

Emadi, M.H. 1998. IK: Sustainability and Empowerment. Indig. Know. & Devt. Monitor. 6(3): 16.

Estelitta, S., Elsy, C.R, and Jayasree, S. 2000. "GALASA for sustenance of rice culture in Kerala". Inter National Conference on Democratic Decentralization, Trivandrum, Kerala State Planning Board.

Eyzaguirre, Pablo, B. 2001. Global Recognition of Indigenous Knowledge: Is this the Latest Phase of 'Globalisation'? Indig. Know & Dev. Monitor. 9 (2): 44.

FAO. 1992. International Conference on Nutrition–Major Issues for Nutrition Strategies.

Fernandez, Maria E. 1994. Gender and Indigenous Knowledge. Indig. Know. & Devt. Monitor, 2(3): 6-7.

Flavier, J.M. 1995. The Regional Program for the Promotion of Indigenous Knowledge in Asia. The Cultural Dimension of Development: Indigenous Knowledge Systems (eds. Warren, D.M., Silkkerveer, E.J. and Brokensha, D.) Intermediate Technology Publications, London, pp. 479-487.

Ganeshamoorthi, S. 2000. Indigenous Knowledge on Post Harvest Practices. Unpub. M. Sc. (Ag.) Thesis, TNAU, Coimbatore.

Garg, M.K. and M.M. Hoda 1978. Role of Technology in Rural Development. International Conference on Strategy for Rural Development. New Delhi.

Gill, D. 1993. O.K; The Data's Lousy, But Its All We've Got! Gate Keeper Series. International Institute of Environment and Development, United Kingdom, 38p.

Gill, G.J. 1991. Indigenous Erosion Control Systems in the mid – hills of Nepal. Proceedings of a workshop on Farmers' practices and Soil and Water Conservation Programmes. ICRISAT, Hyderabad.

Gladwin, C. 1980. Adoption of CIMMYT Fertilizer Recommendations for Plan Puebla. In: Indigenous Knowledge Systems and Development (Eds.) D. Brokensha, D. M. Warren and O. Werner. University Press of America, Lanham, Maryland.

Goldman, A.L. 1991. Traditional change in post harvest pest management in Kenya. Agrc. Hum.Value, 8: 22-23.

Gottreich, M., D. Bradu and Y. Halovy. 1964. A Simple Method for Determining Average Banana Fruit Weight. Ktavim, 14: 161-162. Cited from Lahav, E. 1972. Trop. Agric. (Trinidad), 49: 321-325.

Grenier, L. 1998. Working with Indigenous Knowledge – A Guide for Researchers. Internal Development Research Centre, Ottawa, p. 317.

Gupta, A.K. 1987. Scientific Perception of Farmers Innovation in Dry Regions: Barriers and Scientific Curiosity. Working Paper No. 699, Indian Institute of Management, Ahemadabad, p.36.

Gupta, A.K. 1990. "Documneting Indigenous Farmers' Practices", ILEIA News letter, 6(2): 29-30.

Gupta, A.K. 1994. Development and diffusion of technology: Recasting extension science for dry lands. In dry land farming in India: Constraints and Challenges, ed. J.L. Raina, Jaipur, 1994. An earlier version was presented at the International Conference, 'Making the Link', ISNAR, The Hague, November, 1989.

Gupta, A. 1990. Documenting Farmers innovations: or how do people survive through Innovations in Risky Regions. Unpublished paper. Ahmedabad: Indian Institute of Management.

Gupta, A.K. 2000. Grass Roots Innovations for Survival. LEISA – INDIA Newsletter, 2(2): 20-21.

Gupta, A.K. 2000. Shifting Cultivation and Conservation of Biological Diversity in Tripura, Northeast India. Hum. Ecol. 28: 605-629.

Gupta, A.K. 2002. Rewarding creativity for conserving diversity in third world: can IPR regime serve the needs of contemporary and traditional knowledge experts and communities in third world. Indian fmr. Dig. 33 (1): 15-19.

Hanyani–Mlambo, B.T. and Paul Hebinck 1996. Formal and Informal Knowledge networks in conservation forestry in Zimbabwe. Indig. Know & Dev. Monitor. 4 (3): 3-6.

Haverkort, B. and Zeeuw, M. 1991. "Development of Technology towards sustainable Agriculture". Institutional Implications. Agricultural Extension – World wide Institutional Evolution and Forces for Change (eds. Rivera, W.M. and Gustafun, D.J.). London, United Kingdom pp. 75 - 107.

Haverkort B, van der Kamp J. and Waters-Bayer A. 1991. Joining Farmer's Experiments: Experiences in Participatory Technology Development, Intermediate Technology Publications. p. 281.

Haverkort, B. and H de Zeeuw 1992. "Development of Technologies towards sustainable Agriculture". Institutional Implications. Pp 231-242, In W.M. Rivera and D.J. Gustafson (Eds.), Agricultural Extension: World wide Institutional Evolution and forces for change, New York: Elsevier science publishing company.

Hess, C.G. 1997. Hungry for Hope: On the Cultural and Communicative Dimensions of Development in Highland Educator. Intermediate Technology Publications Ltd., London, p. 113.

Hubert, B. 1991. Comparing two methods for implementing farming system surveys in the Imbo region of Barundi 11th Farming Systems Research and Extension Symposium, 2-4 November, 1991. Michigan State University, Michigan, United States of America. Abstract: 15.

Hunn, E. 1985. The Utilitarian in Folk Biological classification. Directions in Cognitive Anthropology. (ed. Doughery, J.), University of Illionis Press, Urbana, pp. 167-168.

Indian Council of Agricultural Research 1995. Report of the Committee on Partnership, Resource Generation, Training Consultancy, Contract Research/Contract Service and Incentive Research System. Indian Council of Agricultural Research, New Delhi, 72p.

International Institute for Rural Reconstruction. 1996. Recording and Using Indigenous Knowledge: A Manual International Institute of Rural Reconstruction. Siland. Cavite, Philippines.

Joshi, C.P. and B.B. Singh. 2006. Indigenous Agricultural Knowledge in Kumaon Hills of Uttaranchal, Indian Journal of Traditional Knowledge, 5(1): 19-24.

Juma, C. 2000. Agricultural Biotechnology: Risks and Opportunities for Developing Country Food Security. Int. J. Biotech. 2(1): 8 – 10.

Kanagasabapathi, K. 1993. Thriving Indigenous Knowledge in Tamil Nadu. Honey Bee 4 (5): 10 -11.

Kanagasabapathi, K. 1996. Indigenous Knowledge System of Tribals for Agricultural Development. Unpub. Ph.D. Thesis, Dept. of Agrl. Extension, Annamalai University, Annamalainagar.

Karter, A. 1993. "Indigenous Learning in Crafts: A Pilot Research Effort". Indig. Know & Devt. Monitor, 1(1): 20-22.

Karthikeyan, C., Veeraragavathatham, D., Karpagam, D. and S. Ayisha Firdouse 2006. Cow based Indigenous Technologies in Dry Farming, Indian Journal of Traditional Knowledge, 5(1): 47-50.

Kashem, S.K. 1999. Use of Agricultural Technologies by the Rural Men and Women Farmers in Bangladesh. Journal of Sustainable Agriculture 11 (3): 27-43.

Kashyap, S.K., Singh, S.N. and Srivastava, J.P. 2000. Accelerating Rural Development through Indigenous Knowledge System. A Case Study. Indian Fmr. Dig. 33 (3): 23-25.

Khanna, N.P. and Bisa, U.K. 1997. Indian Camel Pastoral Production System and its Indigenous Knowledge. Indian Farming, 47 (3): 28-32.

Kibwana, O.T; Haile, M., Veldhuizen, L. Van and Waters, Bayer, A. 2001. Clapping with two hands: brining together local and outside knowledge for innovation in land husbandry in Tanzania and Ethiopia; a comparative case study. Journal of Agricultural Education and Extension (Netherlands). 7(3): 133-142.

Kieft, J. 2002. Indigenous Variety Development in Food Crops Strategies on Timor: Their Relevance for in situ Bio Diversity Conservations and Food Security. Indian fmr. Dig. 35(1): 20-25.

Kimmerer, R.W. 2000. Native Knowledge for Native Ecosystems, J. Biotech, 98(8): 4-9.

Kishore Kumar, N. and I. Suthan 2004. Vegetable Farmers Perception about PTD and their participation in PTD, Journal of Extension Education, 15(4): 3710 – 3711.

Klemn, S.B. 2000. Biotechnology and Traditional Knowledge; in Search of Equity, Int. J. Biotech. 2: 85-102.

Kothur, S.C. and Keshavamurthy 2007. Enhancement of Bunch Size of Robusta and Ney Poovan Banana, ICAR News, A Science and Technology News Letter, 13(2):1-2 p.

Kshirsagar, K.G. 1991. The Use of Combinations in Reducing Soil Erosion and Degradation. Proceedings of a workshop on Farmers' Practices and Soil and Water Conservation Programmes, ICRISAT, Hyderabad.

Kumar, S. 1994. Co-operative Planning, A Step towards Development of Reservoir Fishermen. Second Congress on Traditional Scientific Technology of India. 27-31 December, 1995, Anna University, Chennai, pp. 38-41.

Kumar, N. and P. Jeyakumar 2001."Influence of Micro Nutrients on Growth and Yield of Banana (*Musa* sp.) cv. Robusta (AAA). W.J. Horst *et al.* (Eds.), Plant nutrition – Food security and Sustainability of Agro – ecosystems. 354 – 355, Kluwer Academic Publishers, Netherlands.

Kumar, A.R. and N. Kumar 2008. Studies on the Efficacy of Sulphate of Potash (SOP) on the Physiological, Yield and Quality Parameters of Banana cv. Robusta (Cavendish – AAA), Eur Asian Journal of Bio Sciences, 2(12): 102 -109.

Kurup, A.M. 2000. Indigenous Knowledge and Intellectual Property Right of Tribals: A Case Study, Yojana, 44(2): 5-9.

Lakshmanan, K.K. 2000. Indigenous Technical Knowledge: A Basis of Sustainable Agriculture. Indian Fmr. Dig. 33 (5): 34-35.

Lal, S. and Singh, A.K. 1997. How to Sustain Agriculture? Indian Fmg, 47 (2): 28-30.

Lawas, Corazon, M. and Henk A. Luning 1996. Farmers Knowledge and GIS. Indig. Know & Devt. Monitor, 4 (1): 8 – 11.

Lilja, N. and Bellon, M.R. (In Press) Participatory Research at CIMMYT. CIMMYT, Mexico, D.F. Mexico.

Lupanga, I.J., Mvena, Z.S.K. and Forster, P.G. 1995. The role of indigenous technical knowledge in increased food production in Tanzania. In Forster. P.G. and Maghimbi, S. 1995 (Eds.) The Tanzanian peasantry further studies: 198-217.

Majumdar, G.P. 1927. Plant and Plant Life in Indian Treatise and Traditions, University of Calcutta, Calcutta, p.143.

Mandal, M.K. and Chauhan, J.P.S. 2001. Awareness of Traditional Wisdom of Animal Feeding Practice. Indian Farming. 34(7): 61-62.

Mane, P.M. and J.N. Sutaria 1993. "Study and Documentation of Indigenous Knowledge/Traditional Agriculture Practice of the Tamil Farmers".

Paper presented at National Seminar on Indigenous Technologies for Sustainable Agriculture, New Delhi, March 23-25, 1993.

Marrewijk. A.V. 1998. Editorial Column. Indig. Know. & Devt. Monitor, 6 (2):1.

Marschke, M. and Nong, K. 2003. Adaptive co-management: Lessons from coastal Cambodia. Canadian journal of development studies (Canada) 24 (3): 369-383.

Martin A. and Sherington, J. 1997. Participatory research methods - Implementation, effectiveness and institutional context. Agric. Syst. 55: 195-216.

Massaquoi, J.G.M. 1993. "Indigenous Technology for Off–Farm Rural Activities". Indig. Know & Devt. Monitor, 1(3): 2-3.

Mendoza, Maria Corazon, Y. 1999. Terracing with the Aid of "Gravity Flow". Indig. Know. & Devt. Monitor. 2(2): 13-17.

Mishra, T.K. 1998. Users become Managers: Indigenous Knowledge and Modern Forestry. Economic and Political Weekly, 33(6): 262 – 263.

Moockes, J.L. and Rhoades, R.E. 1992. Diversity, Farmer Knowledge and Sustainability. Indigenous Know. Dev. Monit. 12(3): 27-29.

Moris, J.R. 1970. Multi–Subject farm Surveys Reconsidered: Some Methodological Lessons. East African Agricultural Economics Society, Dares Salem, 145p.

Morris, M. L. and Bellon, M.R. 2004. Participatory Plant Breeding Research: Opportunities and Challenges for the International Crop improvement system. Euphytica 136: 21 - 35.

National Council of Applied Research. 1993. Comparative Study of Sample Survey and Participatory Rural Appraisal Methodologies with Special Reference to Evaluation of Natural Programme on Improved Chullah. National Council of Applied Economic Research, New Delhi, 154 p.

Nayman, O.B. 1998. Distant cousins: An Evaluation of Official and Non – Official sources of Agricultural Information, Unpublished Manuscript, Punjab, Pakistan.

Neto, E.M.C. 2000. Knowledge and Traditional uses of Animal Resources by an Afro – Boazilian Community. Interciencia. 25: 423-431.

Obinne, C.P. and Ozowa, V.N. 1997. Disseminating Agricultural Information to Rural People Integrating Indigenous Knowledge and Communication, Adult Edu. Dev. 48: 237 – 244.

Okali, C., Sumberg. J. and Farrington, 1997. Farmer Participatory Research: Rhetoric and Reality. Intermediate Publications, London, 159 p.

Okali, C. and Sumberg. J. 1997. Farmers' Experiments: Creating local knowledge, Boulder Colorado: Lynne Rienner Publishers.

Olorunnipa, Z.I. 1993. African Women and Agricultural Technologies. African Technology Forum, 5(1): 25.

Pal, S. Tripp, R. and Janaiah, A. 2000. Public – Private Interface and Information Flow in the Rice Seed System of Andhra Pradesh. National Centre for Agricultural Economics and Policy Research, New Delhi, 41p.

Pamela, G. Fernandez 1994. Indigenous Seed Practices for Sustainable Agriculture. Indig. Know. & Devt. Monitor. 7 (3): 18-19.

Parvathi, S. 1995. A Field Experiment on Farm Women's Cognitive Domain Relating to Post Harvest Technologies. Unpub. Ph.D Thesis, TNAU, Coimbatore.

Parvathi, S., K. Chandrakandan and C. Karthikeyan. 2000. Women and Dry land Post – Harvesting Practices in Tamil Nadu, India. Indig. Know. & Devt. Monitor. 8(1): 13-16.

Pastakia, A.R. 1996. Grassroots innovations for Sustainable Development: The case of agricultural Pest – management. Unpublished fellow programme in Management dissertation. Indian Institute of Management, Ahmedabad.

Paul, S. and Ramanathan, A. 2002. Conservation of Rural Biodiversity through Indigenous Knowledge. Kurukshetra, 50 (3): 3-7.

Pereira, Winin and Anil K. Gupta 1993. A Dialogue on Indigenous Knowledge. Honey Bee, 4(4): 6-10.

Prasad, R. and Power, J.F. 1997. Soil Fertility Management for Sustainable Agriculture, Lewis Publishers, New York, 356. p.

Prasad, R.S., R.P. Singh and R.M. Srivastava 1996. Rationale of Indige-nous Post Harvest Practices in Ranchi District. Journal of Research. 8(2): 187-189.

Preetha, L. 1997. Indigenous Practices in Rice Farming in Thrissur District. Unpublished M.Sc. (Ag.) Thesis, Kerala Agricultural University, Thrissur, India, p.136.

Prema, A. Ahamed, P. and Rehumathniza, T.J. 2000. Validation of Agricultural Technologies through Participatory Research – An Illustrative Case. Proceedings of Twelfth Kerala Science Congress, January 27-29, 2000. (ed. Jayakumar, C.). State Committee on Science and Technology and Environment, Kumily, pp. 780-783.

Pretty, J. and R. Sandbrook 1991. "Operationalising Sustainable Development at the Community Level: Primary Environmental Case". Paper presented at the DAC working party on Development Assistance and the Environment, London, October, 1991.

Prevel, M.P.J. 1989. Physiological Processes Related to Handling and Storage Quality of Crops. In: Methods of K Research in Plants. Proceedings of the 21st Colloquium of the International Potash Institute held at Louvain – la- Neuve, Belgium, 19-21 June 1989. IPI, Bern, Switzerland. Pp. 219-248.

Pulmate, L. and A. Ramesh Babu 1993. "A Reasoned Exposition of the Traditional Farm Practices under Use by the Farmers of Shifting and Settled Cultivation Systems in Manipur". Paper Presented at the

National Seminar on Indigenous Technologies for Sustainable Agriculture, New Delhi, March 2-25, 1993.

Quist, G.D. 1996. Private Commercial Cultivars and Seeds: Opportunities and Obstacles: Overseas Development Institute, London, 135 p.

Rajapakse, W.U.K. 1994. Quoted In: Some Impressions of a Symposium (Eds.) Ulluwishewa, R., D.M. Warren and A. Tick. Indig. Know. & Devt. Monitor. 2(1): 22-24.

Rajaram, G., Erbach, D.C. and Warren, D.M. 1991. The Role of Indigenous Tillage Systems in Sustainable Food Production. Agric. Hum. Values 8 (2): 14-15.

Rajasekaran, B. 1993. A Framework for Incorporating Indigenous Knowledge Systems into Agricultural Research, Extension and NGO's for Sustainable Agricultural Development. Studies in Technology and Social Change No. 21. Technology and Social Change Program, Iowa State University, Iowa, Ames.

Rajasekaran, B. and R.A. Martin 1990. "Evaluating the role of farmers in Training and Visit Extension System in India" Paper presented at the Tenth farming Systems Research/Conference. East Lansing, Michigan, October 14-18, 1990.

Rajasekaran, B., R.A. Martin and D.M. Warren 1993. A Framework for Incorporating Indigenous Knowledge Systems in to Agricultural Extension. Indig. Know. & Devt. Monitor. 2(2): 13-17.

Rajasekaran, B. and M.B. Whiteford 1993. The Role of Indigenous Knowledge in Designing Food Security Policies. Food Policy. 18(3): 237-247.

Rajulasanthy, K. 1991. Indigenous Resource Management by Farm Women – An Analysis. Unpub. M. Sc. (Ag.) Thesis, TNAU, Coimbatore.

Ramaraju, K.V. 1993. Indigenous Technological Knowledge and factors Influencing the Tendency of Tribal and Non-tribal Farmers towards Traditional Practices. Unpub. M.Sc. (Ag.) Thesis, Andhra Pradesh Agricultural University, Hyderabad.

Rambabu, Puli 1997. Indigenous Technologies in Cropping Systems – An Analytical Study in Guntur District of Andhra Pradesh. Unpub. Ph.D. thesis, Acharya N.G. Ranga Agricultural University, Hyderabad.

Ramkumar, S. 1998. Realising the Reality: the First Step in Sustainable Agricultural Development, Agric. Res. Ext. Newsletter, 37 (1): 22-24.

Rani, S., Varama, S.K. and Singh, I. 2002. The Indigenous Animal Health Techniques in Kumaon Hill of Uttaranchal State, Indian Fmr. Dig. 35 (9): 41 - 43.

Rath, S. 1993. "Participatory Research Approach: A strategy for Integrating Local Technical Knowledge with Formal Research System". Paper Presented at the National Seminar on Indigenous Technologies for Sustainable Agriculture. New Delhi. March 23 – 25, 1993.

Ravikumar, R.K., Rao, B.S., Bose, S.C., Sudhakar, K. and Anganappa, M. 2002. Indigenous Technical Knowledge of Livestock Owners in Animal

Husbandry Practices for Managing Genital Distress, Indian Fmg, 35 (7): 39-42.

Reed, M.S. 2007. Participatory technology development for agro forestry extension: an innovation-decision approach African Journal of Agricultural Research 2(8): 334-341.

Reddy, L. Narayana 1988. Experiences with Natural farming. Proceedings of Bio – Control Seminar: pp. 49-54.

Reij C. and Waters-Bayer A. (2001) Entering research and development in land husbandry through farmer innovation. In: Farmer Innovation in Africa: a source of inspiration for agricultural development, Reij C and Waters-Bayer A (eds), Earthscan Publications pp. 3-22.

Reijntjes, C. 1992. Farming for the future: An introduction to low external input and sustainable agriculture. The McMillan Press Ltd., London, P. 293.

Reijntjes, C., Haverkori. B. and Bayer, W.A. 1992. Farming for the future: An introduction to Low External – Input and Sustainable Agriculture. The Macmillan press Ltd., London, P. 423.

Rhoades, Robert and Anthony Bebbington 1994. Framers Who Experiment: An Untapped Resource for Agricultural Research and Development. In: The Cultural Dimensions of Development: Indigenous Knowledge Systems (Eds.) D.M. Warren, L.J. Slikkerveer and D. Brokensha. 1994: pp. 296-307. Intermediate Technology Publications Ltd., London.

Rocheleau, D.E. 2003. Participation in Context: What's Past, What's Present and What's Next. *In:* Pound, B., S. Snapp, C. McDougall and A. Braun. (eds). Managing Natural Resources for Sustainable Livelihoods: Uniting Science and Participation. London: Earthscan and Ottawa: IDRC. pp 169-183.

Roling, N. and P. Engel 1992. The Development of the Concept of Agricultural Knowledge Information Systems (AKIS): Implications for Extension. In: Agricultural Extension Worldwide Institutional Evolution and Forces for Change (Eds.) W.M. Rivers and D.J. Gustafson. Pp. 123 – 138, Elsevier Science Publishing Company, New York.

Rudramoorthy, K. 1964. Extension in Planned Social Change. Allied Publishers Pvt. Ltd, Bombay, p. 296.

Sain, B.C. and Knecht, R.W. 1995. Analysis of Earth Summit Prescriptions on Incorporating Traditional Knowledge in Natural Resource Management. Property Rights and Environment: Social and Ecological Issues (eds. Hanna, S. and Munasinghe, M.), Intermediate Technology Publications Ltd., London, pp. 105-117.

Salas, M.A. 1994. The Technicians only Believe in Science and Cannot Read the Sky the Cultural Dimension of Knowledge Conflict in the Andes. Beyond Farmer First. (eds. Scoones, I. and Thompson, N.). Intermediate Technology Publications Ltd., London, p 271.

Samantha, R.K. 2002. Indigenous Agro technology. Kissan Wld 27 (4): 8-10.

Sanghi, N.K. and J.M. Kerr 1991. The Logic of Recommended and Indigenous Soil and Water Conservation Practices: Proceedings of a workshop on Farmers' Practices and Soil and Water Conservation Programmes, ICRISAT, Hyderabad.

Scoones, I. 1989. Patch Use by Cattle in Dry Land Zimbabwe: Farmer Knowledge and Ecological Theory. ODI Pastoral Development Net Work Paper No. 28. Overseas Development Institute, London.

Seeland, K. 2000. What is Indigenous Knowledge and why does it matter today. Man in the Forest (eds. Seeland, K. and Franz, S.) Printworld, New Delhi, p.775.

Seetharaman, R.N. 2001. Seasonal Forecasting. Short course on capturing the benefits of seasonal climate forecasts in Agricultural Management. TNAU, Coimbatore.

Senthil Vinaygam, Buddheshwar Umraoji Dupare and Om Prakash Joshi 2006. Traditional Technologies in Soybean cultivation in Madhya Pradesh, Indian Journal of Traditional Knowledge, 5(1): 25-33.

Shah, P. 1994. Local Institutions and Para-Professionals in Watershed Management. ILEIA Newsletter, 10 (2): 17.

Sikhana, P. 1994. Alternatives to Current Agricultural Research and Extension System. Intermediate Technology Publication Ltd., London. P. 374.

Singh, A.K., Kunwar, N. and Devi, S. 2001. Proper Storage a Solution of World Hunger. Intensive Agric. 39(2): 6-7.

Singh, C.M., Kharwara, P.C., Singh, V., Sharma, S.K., Prakash, S. and Sharma, D.C. 2001. Study of Indigenous Technological Knowledge for Scientific Use may prove a Boon for Sustaining Agricultural Productivity in Hills. Indian Fmg, 50 (10): 43 - 46.

Somasundaram, S. 1995. Indigenous Knowledge in Farming Systems. Unpub. Ph.D. Thesis, TNAU, Coimbatore.

Sthapit, B.R., Joshi, K.D. and Witcombe, J.R. 1996. Farmer Participatory Plant Breeding: A case study for Rice in Nepal. Exp. Agric. 32: 479 -496.

Stroud, A. 1993. Conducting on-farm experiments. Cali, Columbia: CIAT.

Sulaja, O.R. 1999. Endangered Skills in the Farming System of Mukundapuram Thaluk in Thrissur District. Unpublished M.Sc (Ag) thesis, Kerala Agricultural University, Thrissur, India. P.167.

Sumberg. J., Okali, C. and Reece, D. 2003. Agricultural Research in the face of diversity, local knowledge and the participation imperative: Theoretical considerations. Agricultural – Systems (UK), 76(2): 739 -754.

Sundaramari, M. 2003. Indigenous Agricultural Practices for Sustainable Farming. Agrobios (India). Jodhpur.

Sundaramari. M, Ranganathan, T.T. and R. Netajiseetharaman 2004. Methodology to measure the perceived effectiveness of Indigenous Agricultural Practices, Journal of Extension Education, 15(4): 36-78.

Sunil, G. 1998. A Study of the Traditional Belief System in Diary Husbandry among Tribals of Attappady. Unpublished M.Sc. (Ag.) thesis, Kerala Agricultural University, Thrissur, India. P.172.

Suresh, D. R. and Hegde, G.V. 2002. Indigenous Methods of Treating Seeds for Better Germination, Indian Fmg, 35 (1): 33-34.

Suzuki, D. and Knudtson, P. 1992. Wisdom of the Elders. Toronto: Stoddart Publishing Co.

Swapna, T.R. and P. Ahamed 2005. Scientific Rationality of ITK on Pest Management in Coconut, Indian Coconut Journal, Vol. XXXVI(8): 20-23.

Talawar, S. and Y.P. Singh 1992. Indigenous Knowledge System: Climate and Crop Pest Climate Interactions. Indian Journal of Extension Education, 28 (1,2): 1-7.

Talawar, S. and Y.P. Singh 1993. "In Search of Wisdom among the framers of Arid Agriculture: Documentation and Analysis". Paper Presented at the National Seminar on Indigenous Technologies for Sustainable Agriculture. New Delhi. March 23 – 25, 1993.

Talawar, S. and Y.P. Singh 1994. Understanding Indigenous Knowledge System in Arid Agriculture. Journal of Rural Development, 13 (1): 63-74.

Thiele, G., Gardner, G., Torrez, R. and Gabriel, J. 1997. Farmer Involvement in Selecting New Varieties: Potatoes in Bolivia. Exp. Agric. 33: 275 - 290.

Thurston, H.D. 1990. Plant Disease Management Practices of Traditional Farmers. Plant Disease, 7(92): 96-102.

Titilola, T. 1994. IKS and Sustainable Agricultural Development in Africa: Essential Linkages. Indig. Know. & Devt. Monitor. 2(2): 18-21.

Tripathi, H., Mandape, M.K. and Khandekar, P. 1997. Traditional Veterinary Practices in Northern Plains of Uttar Pradesh. International Conference on Creativity Innovations at Grassroots for Sustainable Resource Management, 8-10 September, 1997, Indian Institute of Management, Ahemadabad, Abstract: 49.

Ulluwishwa, R.1993. Indigenous knowledge national indigenous knowledge resource centres and sustainable development. Indigenous knowledge Development Monitor, 1(3):11-13.

Van veldhuizen, L., A. Waters – Bayer and H. De. Zeeuw 1997. Developing Technologies with farmers: A trainers' guide for participatory learning. London: Zed Books.

Vasu, K.I. 1994. Indigenous Know – how Science Technology and Self – reliance (ed. Ravikumar, K.). State committee on Science, Technology and Environment, Govt. of Kerala, pp.17-21.

Vel, J. 1995. Indigenous Economics: a different rationale ILEIA Newsletter, 1(2): 21-22.

Verma. N.S. and Dhukia, R.S. 1991. Traditional practices of dry land farming based on experiences of generation drawn from solid logics. International conference on extension strategy for minimizing risk in rain fed agriculture. 7-9 August, 1991. Indian Society of Extension Education, New Delhi, Abstract: 19.

Verma, T., Varma, S.K. and Jain, V. 1997. Traditional Medicine and Homest ead Technologies – Need for Scientific Evaluation, Kurukshetra, 25 (2): 48 -49.

Vijayalaxmi, K. 1993. Traditional Indian Agriculture – Potential and Prosp- ects, Second Congress Traditional Sciences and Technology of India, 2-5 February, 1993, PPST Foundation and Indian Institute of Technology, Mumbai, Abstract: 12.

Vijayalaxmi, K. 1996. The Organic Farming. The other India Press, Goa, p.198.

Vivekanandan, P. 1993. "Approaches in Documenting Traditional Technologies. Process and Outcome in Tamil Nadu". Paper Presented at the National Seminar on Indigenous Technologies for Sustainable Agriculture. New Delhi. March 23 – 25, 1993.

Wang, 1988. Indigenous Communication Systems in Research and Development. Journal of Extension System, 4 (2): 75-86.

Wallace – Bruce, Yvonne 1991. Shea Butter Extraction in Ghana. DIH Research Paper presented at the ITDG/UNIFEM Seminar on "Do it herself: Women and Technological Innovation in Africa" in Harare, Zimbabwe.

Warren, D.M. 1987. "Editors Notes" CIKARD News 1 (1): 5.

Warren, D.M. and K. Cashman 1988. Indigenous Knowledge for Agriculture and Rural Development: SomePractical Applications. Paper Presented at the Conference on "Indigenous Knowledge Systems". Academy for Educational Development, Washington, D.C.

Warren, D.M. 1989. Linking Scientific and Indigenous Agricultural Systems. In: The Transformation of International Agricultural Research and Development (Ed.) L.J. Compton: pp. 150–170. Lynne Rienner Publishers, Boulder.

Warren, D.M. 1990. "Indigenous Knowledge Systems and Development", Background Paper for Seminar Series on Sociology and Natural Resource Management. The World Bank, Washington, D.C. December 3, 1990.

Warren, D.M. 1991. Using Indigenous Knowledge for Agricultural Development. World Bank Discussion Paper 127.

Warren, D.M. 1992. International Symposium on Indigenous Knowledge and Sustainable Development. Silang, Philippines. In: Indig Know. & Devt. Monitor, 1(2): 2-5, 1993 and 6(3): 13, 1998.

Warren, D.M. 1992. "Indigenous Knowledge and Sustainable Development: A Review of Critical Research Areas and Policy Issues", Paper presented at the International Symposium on Indigenous Knowledge and Sustainable Development. Silang, Cavite, Philippines, September 20-26, 1992.

Warren, D.M. and Blunt P. 1996. Indigenous Organisations and Development. London: Intermediate Technology Publications.

Weltzien, E., Smith, M.E., L.S. and Sperling, L. 2000. Technical and Institutional Issues in Participatory Plant Breeding from the Perspective of formal Plant Breeding: A Global Analysis of Issues, Results and Current Experience. Consultative group on International Agricultural Research, United States of America, 106p.

Winter, R. 1989. Action Research and the Nature of Social Inquiry: Professional Innovation and Educational Work. Gower Publishing Company, England, 120p.

Witcombe, J.R., Joshi, A. and Joshi, K.D. 1996. Farmer participatory crop improvement, varietal selection and breeding methods and their impact on bio diversity. Exp. Agric. 32: 445-460.

Worade, M. and Mekhib, H. 1993. Looking Genetic Resource Conservation to farmers in Ethiopia. Cultivating Knowledge, Genetic Diversity, Farmer Experimentation and Crop Research (eds. Worade, M. and Mekbib, H.) Intermediate Technology Publications, London, pp. 78 -84.

World Bank. 1997. Knowledge for Development, World Development Report, Oxford University Press, U.K. p. 160.

APPENDIX – 1

Questionnaire for Scientists and Extensionists

REFINEMENT OF IDENTIFIED INDIGENOUS TECHNICAL PRACTICE IN BANANA THROUGH PARTICIPATORY TECHNOLOGY DEVELOPMENT (PTD)

1. Name of the respondents :

2. Age :

3. Sex :

4. Address :

5. Educational Qualifications :

6. Occupation :

7. Degree of Dependability in Treatments in Banana

Treatment number	Treatment details	Degree of confidence		
		Highly Confident (3)	'Some what' Confident (2)	No Confidence at all (1)
T_1	House hold wood ash + Pregnant cow urine			
T_2	Brick Chamber wood ash + Pregnant cow urine			
T_3	Panchakavia + Brick Chamber wood ash			

T_4	Panchakavia + Household wood ash
T_5	Urea
T_6	Urea + Sulphate of Potash
T_7	Fresh cow dung extract + Zinc sulphate + Borax
T_8	Fresh cow dung extract + Zinc sulphate
T_9	Fresh cow dung extract + Borax
T_{10}	Urea + Sulphate of potash + water + Fresh cow dung
T_{11}	Cow dung extract solution
T_{12}	Control

8. Adoption of Treatments in Banana

Treatment number	Treatment details	Adopted		Will be adopted	
		Yes (4)	No (1)	Yes (2)	No (1)
T_1	House hold wood ash + Pregnant cow urine				
T_2	Brick Chamber wood ash + Pregnant cow urine				
T_3	Panchakavia + Brick Chamber wood ash				
T_4	Panchakavia + Household wood ash				
T_5	Urea				
T_6	Urea + Sulphate of Potash				
T_7	Fresh cow dung extract + Zinc sulphate + Borax				
T_8	Fresh cow dung extract + Zinc sulphate				
T_9	Fresh cow dung extract + Borax				
T_{10}	Urea + Sulphate of potash + water + Fresh cow dung				
T_{11}	Cow dung extract solution				
T_{12}	Control				

9. Assignment of Weightage Score to Treatments in Banana

Treatment number	Treatment details	Marks out of 10
T_1	House hold wood ash + Pregnant cow urine	
T_2	Brick Chamber wood ash + Pregnant cow urine	

T_3	Panchakavia + Brick Chamber wood ash
T4	Panchakavia + Household wood ash
T_5	Urea
T_6	Urea + Sulphate of Potash
T_7	Fresh cow dung extract + Zinc sulphate + Borax
T_8	Fresh cow dung extract + Zinc sulphate
T_9	Fresh cow dung extract + Borax
T_{10}	Urea + Sulphate of potash + water + Fresh cow dung
T_{11}	Cow dung extract solution
T_{12}	Control

10. Evaluative Perception of Treatments in Banana

Treatment number	Treatment details	Perceived Effect				
		Most effective (5)	Effective (4)	Undecided (3)	Less effective (2)	Least effective (1)
T_1	House hold wood ash + Pregnant cow urine					
T_2	Brick Chamber wood ash + Pregnant cow urine					
T_3	Panchakavia + Brick Chamber wood ash					
T4	Panchakavia + Household wood ash					
T_5	Urea					
T_6	Urea + Sulphate of Potash					
T_7	Fresh cow dung extract + Zinc sulphate + Borax					
T_8	Fresh cow dung extract + Zinc sulphate					
T_9	Fresh cow dung extract + Borax					
T_{10}	Urea + Sulphate of potash + water + Fresh cow dung					
T_{11}	Cow dung extract solution					
T_{12}	Control					

11. Rationalization of Treatments in Banana

Treatment number	Treatment details	Scientific Rationale				
		Very rational (5)	Rational (4)	Un decided (3)	Irrational (2)	Very irrational (1)
T_1	House hold wood ash + Pregnant cow urine					
T_2	Brick Chamber wood ash + Pregnant cow urine					
T_3	Panchakavia + Brick Chamber wood ash					
T_4	Panchakavia + Household wood ash					
T_5	Urea					
T_6	Urea + Sulphate of Potash					
T_7	Fresh cow dung extract + Zinc sulphate + Borax					
T_8	Fresh cow dung extract + Zinc sulphate					
T_9	Fresh cow dung extract + Borax					
T_{10}	Urea + Sulphate of potash + water + Fresh cow dung					
T_{11}	Cow dung extract solution					
T_{12}	Control					

12. Perception towards Participatory Technology Development (PTD)

Kindly record your response in the four–point continuum Strongly Agree, Agree, Disagree, Strongly Disagree by putting (√) mark in the appropriate column.

Sl. No	Statement	Perception towards PTD			
		Strongly Agree (4)	Agree (3)	Disagree (2)	Strongly Disagree (1)
1	PTD Plays an important role and paves the way to the scientists to expand their imagination about different modes of research				

2 PTD is important for the Researc-hers/ Extensionists to interact with those who operate out side the institution including farmers, community – based organiz-ations, NGOs and Institutions of higher learning.

3 PTD enable the researchers to convince on the superiority of their knowledge as compared to that of farmers and other groups.

4 In PTD, the knowledge is generated both by researchers and farmers, and the researchers have to share knowledge and recognition with farmers.

5 Through PTD farmer participation ends with good research and is a better way for Farmer to Farmer Communication.

6 PTD is to make research more respon-sive to farmer needs and adjustment of the research agenda.

7 Participatory research is the first hand appreciation of the diversity of farmers' problems.

8 Participatory Technology Development is the incorporation of farmers' criteria into technology design and technology evaluation.

9 PTD is a multi disciplinary approach that increases appreciation of socio – economic factors by bio physical scientists.

10 PTD provides identification and use of Indigenous Technical knowledge (ITK) and appreciation for farmer innovation adds value.

11 PTD is a generation of Win – Win techn-ologies using farmer – led experime-ntation

12 PTD is a collaborative activity and synergies between farmers, develo-pment partners and researchers have improved chances for change.

13 PTD can help extensionists to recognize their roles.

14 PTD refers to farmer participatory research where scientists interact with farmers to test and adapt the scientists' ideas.

15 PTD is practiced as an approach to extension with development workers supporting farmers in learning by doing without involving researchers.

16 Extension agents involved in PTD approach live closer to the farmers, use local services and facilities and tailor their work to use resources at hand more efficiently.

17 PTD focusing on local knowledge and resources makes extension agents and farmers equal as partners in their research pursuits.

18 Transfer of Technology paradigm is increasingly replaced by the drive to facilitate PTD

19 Participatory research should involve farmers as co – investigators and researchers and demonstrated through farmer – managed trials.

20 PTD is frequently used as a final stage in technology development.

21 PTD plays an effective role on confronting the tendencies of academics to look down on farmers' indigenous knowledge.

22 PTD advocates and stress the importance of organizing farmers into groups.

23 PTD indicates that the researchers need to have the ability to engage in dialogue, and listening rather lecturing.

24 PTD is an emerging approach that does not deal with problem analysis alone, but rather seeks to link up with local problem – solving initiatives.

Signature of the Respondent

APPENDIX – 2

QUESTIONNAIRE FOR FARMERS

REFINEMENT OF IDENTIFIED INDIGENOUS TECHNICAL PRACTICE IN BANANA THROUGH PARTICIPATORY TECHNOLOGY DEVELOPMENT (PTD)

1. Name of the respondents :

2. Age :

3. Sex :

4. Address :

5. Educational Qualifications :

6. Occupation :

7. Degree of Belief in Treatments in Banana

Treatment number	Treatment details	Degree of belief		
		Strong belief (3)	'Some what' belief (2)	No belief at all (1)
T_1	House hold wood ash + Pregnant cow urine			
T_2	Brick Chamber wood ash + Pregnant cow urine			

T_3	Panchakavia + Brick Chamber wood ash
T_4	Panchakavia + Household wood ash
T_5	Urea
T_6	Urea + Sulphate of Potash
T_7	Fresh cow dung extract + Zinc sulphate + Borax
T_8	Fresh cow dung extract + Zinc sulphate
T_9	Fresh cow dung extract + Borax
T_{10}	Urea + Sulphate of potash + water + Fresh cow dung
T_{11}	Cow dung extract solution
T_{12}	Control

8. Strength of Treatments in Banana

Treatment number	Treatment details	Done		Will Do	
		Yes (4)	No (1)	Yes (2)	No (1)
T_1	House hold wood ash + Pregnant cow urine				
T_2	Brick Chamber wood ash + Pregnant cow urine				
T_3	Panchakavia + Brick Chamber wood ash				
T_4	Panchakavia + Household wood ash				
T_5	Urea				
T_6	Urea + Sulphate of Potash				
T_7	Fresh cow dung extract + Zinc sulphate + Borax				
T_8	Fresh cow dung extract + Zinc sulphate				
T_9	Fresh cow dung extract + Borax				
T_{10}	Urea + Sulphate of potash + water + Fresh cow dung				
T_{11}	Cow dung extract solution				
T_{12}	Control				

9. Assignment of Weightage to Treatments in Banana

Treatment number	Treatment details	Marks out of 10
T_1	House hold wood ash + Pregnant cow urine	
T_2	Brick Chamber wood ash + Pregnant cow urine	

T_3	Panchakavia + Brick Chamber wood ash
T_4	Panchakavia + Household wood ash
T_5	Urea
T_6	Urea + Sulphate of Potash
T_7	Fresh cow dung extract + Zinc sulphate + Borax
T_8	Fresh cow dung extract + Zinc sulphate
T_9	Fresh cow dung extract + Borax
T_{10}	Urea + Sulphate of potash + water + Fresh cow dung
T_{11}	Cow dung extract solution
T_{12}	Control

10. Evaluative Perception of Treatments in Banana

Treatment number	Treatment details	Perceived Effect				
		Most effective (5)	Effective (4)	Un decided (3)	Less effective (2)	Least effective (1)
T_1	Household wood ash + Pregnant cow urine					
T_2	Brick Chamber wood ash + Pregnant cow urine					
T_3	Panchakavia + Brick Chamber wood ash					
T_4	Panchakavia + Household wood ash					
T_5	Urea					
T_6	Urea + Sulphate of Potash					
T_7	Fresh cow dung extract + Zinc sulphate + Borax					
T_8	Fresh cow dung extract + Zinc sulphate					
T_9	Fresh cow dung extract + Borax					
T_{10}	Urea + Sulphate of potash + water + Fresh cow dung					
T_{11}	Cow dung extract solution					
T_{12}	Control					

11. Rationalization of Treatments in Banana

Treatment number	Treatment details	Scientific Rationale				
		Very rational (5)	Rational (4)	Undecided (3)	Irrational (2)	Very irrational (1)
T_1	Household wood ash + Pregnant cow urine					
T_2	Brick Chamber wood ash + Pregnant cow urine					
T_3	Panchakavia + Brick Chamber wood ash					
T_4	Panchakavia + Household wood ash					
T_5	Urea					
T_6	Urea + Sulphate of Potash					
T_7	Fresh cow dung extract + Zinc sulphate + Borax					
T_8	Fresh cow dung extract + Zinc sulphate					
T_9	Fresh cow dung extract + Borax					
T_{10}	Urea + Sulphate of po-tash + water + Fresh cow dung					
T_{11}	Cow dung extract solution					
T_{12}	Control					

12. Perception towards Participatory Technology Development (PTD)

Kindly record your response in the four–point continuum Strongly Agree, Agree, Dis agree, Strongly Disagree by putting (√) mark in the appropriate column.

Sl. No	Statement	Perception towards PTD			
		Strongly Agree (4)	Agree (3)	Disagree (2)	Strongly Disagree (1)
1	PTD became a fully accepted approach for information exchange between farmer and researchers				
2	PTD is the practical process of bringing together the knowledge and research capacities of the local farming communities.				

3 In PTD approach, the vital local knowledge and local people is recognized and tapped in development

4 PTD must be built on effective communication with local groups and on constructive use of local institutions.

5 Researchers and farmers each have to contribute knowledge in PTD.

6 PTD takes advantage of the ability of farmers to experiment and solve problems on their own farms

7 PTD improves the chances of wider adoption of agricultural technologies than technology developed solely on research stations for extension to farmers.

8 PTD is based on active, decision making involvement of farmers in all stages of the technology development process.

9 PTD is an on going (but not endless) process that starts with only a superficial understanding of technologies that may solve farmers' problems

10 PTD process experimenting locally with a variety of options derived both from indigenous knowledge and from formal science.

11 PTD arouses the farmers' inherent curiosity and creativity and increases confidence to continue experimentation

12 As an informal process, PTD results in farmer- to-farmer networking that rapidly and efficiently spreads the ideas.

13 PTD is a cost effective approach as it does not require highly paid scientists.

14 PTD is the centrality of partnership in mutual and respectful learning among farmers, scientists and extension workers

15 PTD involves the farmers in all stages of the development process

16 PTD focuses on strengthening local capacities to experiment.

17 PTD create links between different types of knowledge as well as between different sources of knowledge.

18	PTD activities depend on collaboration among farmers, field workers, researchers and others.
19	As a result of PTD, farmer's innovative capacities are strengthened.
20	PTD strengthens farmer's organizations and other stake holder groups.
21	PTD applications have the expansion of spheres beyond agriculture into marketing and other endeavors.
22	PTD has been advocated as a way of increasing the livelihood that technologies developed will be suitable for resource poor households.
23	PTD can be sustained even without external agencies through a system of farmer experimenters.
24	PTD processes help to bring stakeholders together for stakeholder analysis, roles and task identification.
25	Innovations whether initiated by local people or by scientists are entry points to PTD.
26	The effective use of farmer to farmer extension is a key element in the PTD approaches.
27	PTD plays an effective role to emerge the farmer's local ideas in a successive way.
28	In PTD, there is a wide scope of satisfaction among farmers that their views/ ideas are taken into consideration for evolving a technology.
29	The scientific rationales behind the indigenous technologies are realized by the farmers through PTD approach.
30	PTD paves a way to act as a researcher in the field by the farmers.
31	In PTD, there is a possibility for evolving a low cost/no cost technologies as per farmer's point of view.
32	In participatory research, there is an ample chance of considering the Agro – climatic conditions in the research.

Index